Air Dominance Through Machine Learning

A Preliminary Exploration of Artificial Intelligence–Assisted Mission Planning

LI ANG ZHANG, JIA XU, DARA GOLD, JEFF HAGEN, AJAY K. KOCHHAR, ANDREW J. LOHN, OSONDE A. OSOBA

For more information on this publication, visit www.rand.org/t/RR4311

Library of Congress Cataloging-in-Publication Data is available for this publication.
ISBN: 978-1-9774-0515-9

Support RAND
Make a tax-deductible charitable contribution at
www.rand.org/giving/contribute

www.rand.org

Preface

This project prototypes a proof of concept artificial intelligence (AI) system to help develop and evaluate new concepts of operations in the air domain. Specifically, we deploy contemporary statistical learning techniques to train computational air combat planning agents in an operationally relevant simulation environment. The goal is to exploit AI systems' ability to learn through replay at scale, generalize from experience, and improve over repetitions to accelerate and enrich operational concept development. The test problem is one of simplified strike mission planning: given a group of unmanned aerial vehicles with different sensor, weapon, decoy, and electronic warfare payloads, we must find ways to employ the vehicles against an isolated air defense system. Although the proposed test problem is in the air domain, we expect analogous techniques with modifications to be applicable to other operational problems and domains. We discovered that leveraging the proximal policy optimization technique sufficiently trained a neural network to act as a combat planning agent over a set of changing complex scenarios.

Funding

Funding for this venture was made possible by the independent research and development provisions of RAND's contracts for the operation of its U.S. Department of Defense federally funded research and development centers.

Contents

Figures

Tables

Summary

U.S. air superiority, long a cornerstone of the American way of war and deterrence, is increasingly challenged by developments from competitors. The democratization of machine learning (ML) provides innumerable opportunities for disruption. One potential approach to improve warfighting and reinforce deterrence is through more-effective use of automation, which, in turn, might enable new approaches to planning and development of concepts of operations (CONOPs). By building a ground-level understanding of the strengths and weaknesses of artificial intelligence (AI)–assisted planning through firsthand experimentation, the U.S. defense community will be in a better position to prepare for strategic surprise and disruption.

In this report, we prototype a proof of concept AI system to help develop and evaluate new CONOPs for the air domain. We tested several learning techniques and algorithms to train air-combat agents that can play in operational-relevant simulation environments. The goal is to exploit AI systems' ability to play repeatedly at scale, generalize from experience, and improve over repetitions to accelerate and enrich operational concept development.

To accomplish this task, we integrated open-source deep learning frameworks with the Advanced Framework for Simulation, Integration, and Modeling (AFSIM)—a U.S. Department of Defense–standard combat simulation tool. AFSIM provided the environment for the ML agents to learn. The deep learning frameworks provided the platform to test a myriad of state-of-the-art learning algorithms: generative adversarial networks (GANs), Q-learning, Asynchronous Advantage Actor Critic (A3C), and Proximal Policy Optimization (PPO).

What We Conclude

- Planning on simple toy problems, as exemplified in our one-dimensional missions, was straightforward. Both Q-learning and GANs were successful in generating successful mission plans.
- Two-dimensional (2-D) mission planning was significantly more difficult because learning algorithms had to contend with continuous action spaces (how much to move, how much to turn) and timing strategies (synergize with jammer or decoy effects).
- We developed a fast, low-fidelity version of AFSIM that we called AFGYM: The difficulty of the 2-D missions required the creation of a low-fidelity version of AFSIM for faster environment simulation because training required thousands of simulations.
- Current reinforcement learning algorithms appear to be difficult to implement in practice. Algorithms require hand-tuned hyperparameters. Algorithms are also prone to suffering learning collapse, in which the neural network gets stuck in a optima that outputs nonsensical action values. Only the PPO algorithm provided working strategies in 2-D missions.

What We Recommend

- Although mission planning is a difficult problem, we showed some success in using AI algorithms to develop time- and space-coordinated flight routes. Such algorithms, when approached at scale and with better tuning, might have utility for the U.S. Department of Defense.
- The computational power and time requirements to develop such algorithms against realistic threats are unclear. If training is ever complete, AI mission planners have a clear advantage over existing human or automated planning techniques: speed.
- Reward functions drastically change AI behavior in simulations and often in unexpected ways. Care must be taken in designing such functions to accurately capture risk and intent.
- Real-world mission data are scarce compared with the volumes of data used to train contemporary AI systems. Such simulation engines as AFSIM will be used to develop plans. Algorithms trained to perfection against a simulation might behave poorly in the real world (as seen with autonomous vehicles). Advances in testing and algorithm verifiability is likely necessary before committing to AI solutions for security problems.

Acknowledgments

We are grateful for the support of RAND Ventures for this exploratory research effort. Specifically, we thank Howard Shatz and Susan Marquis for their support of the project and for helping to refine the proposal. We also thank Ted Harshberger and Jim Chow for their oversight and support of the project's execution.

Abbreviations

1-D	one-dimensional
2-D	two-dimensional
3-D	three-dimensional
A3C	Asynchronous Advantage Actor Critic
AFSIM	Advanced Framework for Simulation, Integration, and Modeling
AI	artificial intelligence
CGAN	conditional generative adversarial network
CONOP	concept of operations
CPU	central processing unit
DoD	U.S. Department of Defense
DOTA	Defense of The Ancients (game)
EW	electronic warfare
GAN	generative adversarial network
KL	Kullback-Leibler
MDP	Markov decision process
ML	machine learning
PPO	Proximal Policy Optimization
RL	reinforcement learning
SAM	surface-to-air missile
SEAD	suppression of enemy air defenses
TRPO	Trust Region Policy Optimization
UAV	unmanned aerial vehicle
V&V	verification and validation
VM	virtual machine

1. Introduction

U.S. air superiority—long a cornerstone of the American way of war and deterrence—is increasingly challenged by developments from competitors. Concurrently, the democratization of machine learning (ML) provides innumerable opportunities for disruption. China, for example, is evolving its military to wage "intelligentized" war through strategic investments in artificial intelligence (AI).[1]

One potential approach to improve warfighting and reinforce deterrence is through more-effective use of automation, which, in turn, might enable new approaches to planning and concept of operations (CONOP) development. By building a ground-level understanding of the strengths and weaknesses of AI-assisted planning through firsthand experimentation, the U.S. defense community will be in a better position to prepare for strategic surprise and disruption.

The objective of this project is to prototype a proof-of-concept AI system to help develop and evaluate new CONOPs for the air domain. Specifically, we deploy contemporary generative adversarial networks (GANs) and deep reinforcement learning (RL) techniques to train air-combat agents that can play in operational-relevant simulation environments. The goal is to exploit AI systems' ability to play repeatedly at scale, generalize from experience, and improve over repetitions to accelerate and enrich operational concept development.

The prototype platform integrates open-source deep learning frameworks, contemporary algorithms, and the Advanced Framework for Simulation, Integration, and Modeling (AFSIM)—a U.S. Department of Defense (DoD)–standard combat simulation tool. AFSIM provides both the simulated environment and evaluator. This model is used as the "reality" in which the ML agents learn. The test problem is one of simplified strike mission planning: Given a group of unmanned aerial vehicles (UAVs) with different sensor, weapon, decoy, and electronic warfare (EW) payloads, the ML agents must find ways to employ the vehicles against air defenses. This problem is challenging for at least two reasons:

1. The decision space of the optimal control problem is vast—encompassing the vehicle routes, kinematics, and sequencing, and sensor and weapon tactics and strategies at every time-step of the simulation.
2. Agent actions can have both immediate and long-term, delayed effects. The agent must strike the right balance between near- and far-term utilities.

Although the proposed test problem is in the air domain, we expect analogous techniques to be applicable to other simulation environments. All of the problems addressed here can be solved using combinations of other means, such as simple range and timing calculations, rules of thumb

---

[1] Elsa Kania, "AlphaGo and Beyond: The Chinese Military Looks to Future 'Intelligentized' Warfare," blog post, *Lawfare*, June 5, 2017.

for various threat situations, or risk-minimizing route planners. However, in the longer term, the techniques used here could be much more scalable than these other approaches, all of which also rely on simple scenarios or are computationally intensive.

Motivation

DoD is exploring opportunities to incorporate autonomy, AI, and human-machine teaming into its weapons and operations. Whether as data-mining tools for intelligence analysts, decision aids for planners, or enablers for autonomous vehicle operations, these systems have the potential to provide more accuracy, speed, and agility than traditional tools.[2] Yet operational AI and cognitive autonomy systems also face steep challenges in operator trust and acceptance, computational efficiency, verification and validation (V&V), robustness and resilience to adversarial attack, and human-machine interface and decision explainability. Understanding the potential and limitations of AI-based planning and differentiating reality from hype around AI systems are both vital.[3]

Specifically, as DoD begins to develop and procure more-complex, -distributed, and -connective manned and unmanned "system-of-systems" architectures, it will need new analytic tools to augment and scale operational concept development and evaluation. For example, U.S. Air Force planners might want to know how best to mission plan and employ a group of collaborating and heterogeneous UAVs fitted with complementary sensing, jamming, or strike payloads. A planning system would have to answer such questions as: How many vehicles should have which type of payload? What should the relative spacing and timing between vehicles be? Where should they fly relative to manned aircraft? These are the types of questions we anticipate our software could help address.

The Defense Science Board, in its 2015 *Summer Study on Autonomy*, identified "autonomous air operations planning" and autonomy-enabled "swarm operations" as priority research areas.[4] The Air Force, in its 2016 Autonomous Horizons vision, also called out "assistance in mission planning, re-planning, monitoring, and coordination activities" as a key autonomy application.[5] Yet despite the widespread interest in AI/ML technologies for complex defense planning, the extent and complexity of AI system development needed to solve meaningful air operation

---

[2] For examples of these efforts, see Adam Frisk, "What is Project Maven? The Pentagon AI Project Google Employees Want Out Of," *Global News*, April 5, 2018; and Aaron Mehta, "DoD Stands Up Its Artificial Intelligence Hub," blog post, *C4ISRNET*, June 29, 2018.

[3] For a few examples of setting appropriate expectations, see Jory Heckman, "Artificial Intelligence vs. 'Snake Oil:' Defense Agencies Taking Cautious Approach Toward Tech," blog post, *Federal News Network*, December 12, 2018.

[4] Defense Science Board, *Report of the Defense Science Board Summer Study on Autonomy*, Washington, D.C.: Office of the Under Secretary of Defense for Acquisition, Technology, and Logistics, June 2016.

[5] U.S. Air Force, Office of the Chief Scientist, *Autonomous Horizons: System Autonomy in the Air Force—A Path to the Future*, Vol. 1: *Human Autonomy Teaming*, Washington, D.C., 2015.

planning problems is unknown. This limitation stems—at least in part—from the lack of formal means to characterize problem complexity relative to contemporary connectionist AI systems. The general undertheorization of deep learning techniques and a community bias toward empirical research programs mean that it is difficult to determine a priori if contemporary algorithms can solve a given complex planning problem. This report aims to bridge this gap through pathfinding experimentation—to test whether we can deploy contemporary "second-wave" AI/ML techniques to air mission planning.[6]

Continued advances in autonomy, sensor networks, and datalinks mean that the complexities of air operations are likely to grow much faster than our human ability to reason about them. Because of the complexity of planning today, and the possibility of much greater challenges in the future, this project was intended to explore ML approaches for mission planning. The key research questions include:

- Could contemporary ML agents be trained to effectively exhibit intelligent mission-planning behaviors without requiring training data on billions of possible combinations of situations?
- Could machine agents learn, over repeated play, that jammers need to get close enough to surface-to-air missiles (SAMs) to affect them, but not so close that they get shot down? Could they learn that they need to arrive before the strike aircraft does? Could they learn how many decoys are necessary—and at what approach path—to distract a SAM from a striker?
- Could we build sufficiently generalizable state- and action-space representations to capture the richness of the planning problem? Would the lessons learned generalize robustly across changes in threat location, type, and number?

Background

Mission Planning Today

Military aircraft have traditionally not operated alone, but in "packages" of aircraft designed to support each other. At the minimum, a package might consist of two aircraft—a flight lead and wingman. At the other extreme, a larger package could include strike, suppression of enemy air defense (SEAD), air-to-air, reconnaissance, and jamming aircraft that total in the dozens of aircraft.[7] These different types of capabilities need to be carefully coordinated in time and space so that, for example, strike aircraft are not striking the same target at the same time or are not approaching the target when no defense suppression fighters are nearby to protect them. They also need to be carefully planned against expected threats to avoid leaving some aircraft

---

[6] For a definition of AI "waves," see Defense Advanced Research Projects Agency, "DARPA Announces $2 Billion Campaign to Develop Next Wave of AI Technologies," webpage, September 7, 2018.

[7] Jerome V. Martin, *Victory From Above: Air Power Theory and the Conduct of Operations Desert Shield and Desert Storm*, Miami, Fla.: University Press of the Pacific, 2002.

undefended or without sensor support, operating too close to dangerous areas, or wasting typically scarce support assets. Considering that different types of aircraft are likely operating from different locations, fly at different speeds and altitudes, and include pilots of very different experience levels who never might have flown together, the difficulty level can be appreciated.

To manage all this complexity, mission-package planning is one of the most advanced topics of pilot training. Only pilots with years in the cockpit are responsible for these tasks, and planning and integrating such packages are a major topic of such exercises as Red Flag and advanced training opportunities like the Air Force Weapons School.[8] The approach typically involves first identifying the number and type of reconnaissance and strike assets needed to find and destroy the planned target, then identifying key threats to that package based on their location relative to the target and the planned route of the strike aircraft, the threat's range, speed, and likely level of warning. Once the key threats are known, defenses against them can be planned, including jamming to lower their detection ranges, and dedicated air-to-ground and air-to-air aircraft to intercept SAMs or airborne interceptors, respectively. Each of these defensive missions, in turn, can be a package that needs more-detailed planning. Finally, routes and timing are fine-tuned to ensure safety and appropriate levels of mutual support. With sufficient time, the most-complex packages might be rehearsed with live aircraft or in networked simulators.

However, as aircraft and defenses network are increasing in capability, integration, automation, and speed, this type of labor-intensive manual planning might simply be infeasible. Some of the most complex missions, such as those involving stealth aircraft like the B-2, can require many days and hundreds of people to plan.[9] In a war with an advanced adversary, this amount of time and personnel simply might not be available, especially if many missions are needed every day. Of additional concern are upcoming technologies like unmanned aircraft with very flexible payloads, ability to take additional risk, and perhaps autonomous behaviors.[10] Could human planners, with only a few hours (or, indeed, even shorter available timelines), hope to effectively take advantage of the exponentially larger numbers of possible combinations of manned and unmanned aircraft capabilities and tactics, to say nothing of routing and timing options?

It is important to distinguish between mission planning and route planning. Here we use the term *mission planning* to include not only parameters like where the aircraft will fly, but also how many of what type of aircraft and their relative timing. Route planning in a static,

---

[8] For examples, see U.S. Air Force, Exercise Plan 80: RED FLAG—NELLIS (RF-N), COMACC EXPLAN 8030, ACC/A3O, July 2016a; and U.S. Air Force, F-16 Pilot Training Task List, ACC/A3TO, June 2016b, Not available to the general public.

[9] Scott Canon, "Inside a B-2 Mission: How to Bomb Just About Anywhere from Missouri," *Kansas City Star*, April 6, 2017.

[10] For example, see the Defense Advanced Research Projects Agency System of Systems Integration Technology and Experimentation program (Jimmy Jones, "System of Systems Integration Technology and Experimentation (SoSITE)," blog post, Defense Advanced Research Projects Agency, undated).

noninteractive environment with simple constraints—in which we primarily consider where to fly—is a well-understood problem with efficient solutions, such as Dijkstra's algorithm or A∗ search.[11] These algorithms, akin to commercial navigation and routing systems, typically look to minimize such key variables as route length and time. In national security applications, they can be deployed to control risk. They can include many constraints, such as vehicle performance, keep-out regions, and sensor or weapon range. DoD already uses these tools to help plan missions for individual or small groups of aircraft, typically integrated into the Joint Mission Planning System.[12] Much current work on these types of systems is focused on speed, to handle large-scale problems and support dynamic replanning during a mission when conditions change (for example, a threat appears), and to include additional variables and constraints (e.g., winds, vehicle roll angle constraints, higher-fidelity calculations of objective functions like risk).

Mission analysis and planning is often approached through intuition and heuristics as a means to control the size of the search problem. Although heuristics can help us identify point solutions, these solutions are rarely scalable or reliable enough to consistently evaluate the vast numbers of policy alternatives that arise. Intuition can also fail in high-dimensional problems with large numbers of players and complex sensors and weapon interactions, the very characteristics that might define the future of air warfare.

Little work seems to be underway in the defense or commercial world on automated tools for what we term mission planning—what types and numbers of assets should be used together, and how to achieve an objective.[13] As we will discuss later in the report, as the interactivity among ML agents and between agents and the environment increase, it becomes increasingly difficult to separate the mission planning from the routing problem; specifically, how best to coordinate collaborative platforms becomes a strong function of their location relative to each other and to threats and the relative timing of their effects. Our algorithms aim to integrate both route- and mission-planning functions.

---

[11] Two of many surveys of routing algorithms include C. Goerzen, Z. Kong, and Berenice F. Mettler May, "A Survey of Motion Planning Algorithms from the Perspective of Autonomous UAV Guidance," *Journal of Intelligent and Robotic Systems*, Vol. 57, No. 1–4, January 2010; and Ling Qiu, Wen-Jing Hsu, Shell-Ying Huang, and Han Wang, "Scheduling and Routing Algorithms for AGVs: A Survey," *International Journal of Production Research*, Vol. 40, No. 3, 2002.

[12] For some history on this program, see Mark A. Gillott, *Breaking the Mission Planning Bottleneck: A New Paradigm*, Maxwell Air Force Base, Ala.: Air Command and Staff College, AU/ACSC/099/1998-04, April 1, 1998; and Samuel G. White III, *Requirements for Common Bomber Mission Planning Environment*, Wright-Patterson Air Force Base, Ohio: Department of the Air Force, Air University, Air Force Institute of Technology, AFIT/IC4/ENG/06-08, June 2006.

[13] Government organizations, such as the Defense Advanced Research Projects Agency, recognize these challenges and sponsor projects such as Distributed Battle Management to address them.

An Operational Level Simulation Environment: AFSIM

AFSIM, which we use as the environment to evaluate the effectiveness of our ML work, is a contemporary mission-level simulation tool.[14] Originally conceived as an alternative to the Air Force's existing Integrated Air Defense Systems models, many in the DoD community now see AFSIM as the standard mission-level simulation tool.

AFSIM is object-oriented and agent-based: Individual platforms can be designed with sensors, weapons, track managers, communications networks, and processors (customized or chosen from AFSIM's libraries). The architecture includes all of the tools needed to build and script complex platforms and their interactions: simulation objects; weapons engagement and communications modeling; a customized scripting language; and terrain options, including subsurface and space. Platform behavior is scripted by the user to include routing, firing triggers, imaging and communications guidelines, and other engagement rules. Although platforms interact during the scenario run, the user has the option to output records of any events that take place at any time step, including track detection, firing, and jamming. The built-in "scriptability" and instrumentation of the tool provide a pathway to batch processing and Markov decision process (MDP) formalizations.

MDPs are mathematical formulizations of sequential decisionmaking where actions yield stochastic rewards (either immediate or delayed). MDPs provide the underlying theory behind most reinforcement learning algorithms in which the goal is to maximize the expected rewards over time. Figure 1.1 shows a schematic of a typical reinforcement learning problem. An agent is interacting with an environment, and for every state it finds itself in, the agent has to make decisions (take actions) to reap rewards. Mathematical formulation of such concepts as MDPs are provided for the reader's convenience in Appendix D.

---

[14] Peter D. Clive, Jeffrey A. Johnson, Michael J. Moss, James M. Zeh, Brian M. Birkmire, and Douglas D. Hodson, "Advanced Framework for Simulation, Integration and Modeling (AFSIM)," *Proceedings of the International Conference on Scientific Computing*, 2015.

Figure 1.1. Visual Representation of Reinforcement Learning MDP

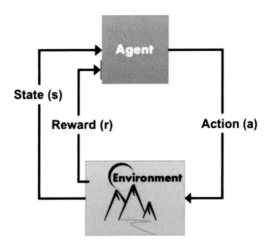

State of Machine Learning in Games

For decades, games have been the pinnacle of AI research, motivating marquee results. Games are the natural laboratory for AI research for several reasons. A key consideration is that games operate in constrained and measurable environments with clearly defined and specified rules and actions. Computers excel when they can be told exactly what to do, but struggle as aspects of the tasks become harder for a programmer to specify. In many games, the rules, actions, and—most importantly—conditions for winning and losing can all be specified with precision, leaving only the strategies partly or completely unspecified. For most real-world problems, few or none of those elements can be so precisely specified. As a result, real-world problems rarely make good test beds for developing AI technologies. However, the same reasons that games can be good playgrounds for AI development also limit the utility of the outcomes, because the goal is rarely to develop AI solely for game playing. The investment in gaming is justified by aspirations for transitioning that technology to solve problems of real-world interest, an outcome that is by no means guaranteed.

That transition process has, at times, been imperiled, in part because the jump directly from the highly prescribed and reduced game environment to the unbounded real-world environment is large. The challenges posed by the divide between games (and simulations) and reality is compounded by the comparative fragility of contemporary ML techniques to even small differences in training and evaluation data. It is instructive, therefore, to review some of the marquee achievements of AI in games, both historical and recent, and some of the struggles and successes in transitioning these achievements out of the "laboratory" environment into more-useful applications.

Perhaps the most famous of all game-playing AIs is Deep Blue, the IBM supercomputer that beat Gary Kasparov in a chess match in 1996.[15] That event marked a turning point for many: the

---

[15] Bruce Weber, "Swift and Slashing, Computer Topples Kasparov," *New York Times*, May 12, 1997.

time when computers overtook humans at one of the premier tests of intelligence and strategy. Leading AI thinkers had believed that a chess-playing computer would be a stepping stone to solving a wide variety of challenging real-world problems, such as "translating from one language to another" and "strategic decisions in simplified military contexts."[16] However, it soon became clear that chess is just a game and that a computer built for the sole purpose of playing chess offers little in the way of extensible utility. Decades later, translation has made impressive progress, but not from techniques closely tied to those used in Deep Blue.[17] Military strategy and planning are still solidly the domain of humans.

More recently, IBM bested humans in another game of intellect, the Jeopardy! game show. The victory was widely heralded, leading 74-time winner Ken Jennings to quip, "I, for one, welcome our new computer overlords."[18] Watson, the winning system, was designed to play this particular game and content with the variability and subtlety in the questions. Many believed that a computer that is able to interpret the questions in this game would excel at a wide variety of practical problems, and Watson was billed as being able to do everything from making music to curing cancer.[19] To be fair, Watson has had some successes in a wide variety of areas, but not, perhaps, commensurate with the scale of ambitions and investment.[20] Once again, it turned out that Watson was largely a purpose-built game-playing program, so transition to real-world problems would require a large degree of redesign. Furthermore, in the real-world environments to which it is trying to transition, Watson faces shortcomings in data availability and quality, as well as variability from person to person and from application to application.[21]

An arguably more-general approach to creating game-playing programs is through RL. The principle behind RL is that the state of a game is provided as inputs to the program, which then suggests the next move based on having seen many previous moves and their eventual outcomes. These programs are able to play only the specific game that they have been trained to play, but the approach and setup do not require programming anything specific to the game, even the

---

[16] Claude E. Shannon, "Programming a Computer for Playing Chess," *Philosophical Magazine*, Vol. 41, No. 314, 1950.

[17] Yonghui Wu, Mike Schuster, Zhifeng Chen, Quoc V. Le, Mohammad Norouzi, Wolfgang Macherey, Maxim Krikun, Yuan Cao, Qin Gao, Klaus Macherey, Jeff Klingner, Apurva Shah, Melvin Johnson, Xiaobing Liu, Łukasz Kaiser, Stephan Gouws, Yoshikiyo Kato, Taku Kudo, Hideto Kazawa, Keith Stevens, George Kurian, Nishant Patil, Wei Wang, Cliff Young, Jason Smith, Jason Riesa, Alex Rudnick, Oriol Vinyals, Greg Corrado, Macduff Hughes, and Jeffrey Dean, "Google's Neural Machine Translation System: Bridging the Gap Between Human and Machine Translation," *arXiv*:1609.08144, last updated October 8, 2016.

[18] John Markoff, "Computer Wins on 'Jeopardy!': Trivial, It's Not," *New York Times*, February 16, 2011.

[19] Marcus Gilmer, "IBM's Watson Is Making Music, One Step Closer to Taking Over the World," *Mashable*, October 24, 2016; Maja Tarateta, "After Winning Jeopardy, IBM's Watson Takes on Cancer, Diabetes," *Fox Business*, October 7, 2016.

[20] Jennings Brown, "Why Everyone Is Hating on IBM Watson—Including the People Who Helped Make It," *Gizmodo*, August 10, 2017.

[21] David H. Freedman, "A Reality Check for IBM's AI Ambitions," blog post, *MIT Technology Review*, June 27, 2017.

rules, so it is a degree more generalizable than chess or question-answer games. As a testament to this generalizability, the same design was shown moves and outcomes for 49 different classic Atari video games and was rapidly able to demonstrate performance comparable to a professional human game tester.[22]

A similar approach was used to outperform the world champion in the ancient Chinese strategy game Go, in a competition reminiscent of the Deep Blue versus Kasparov matches of the 1990s.[23] The program, called AlphaGo, uses the same basic approach as the Atari-playing program, albeit with a different neural network architecture that is more aligned with the specifics of Go. Yet the same basic algorithm and architecture could be applied to this new game or any number of other games, at presumably reduced performance. That transferability has tempted many industries into believing that transitioning from games to real-world applications might be achievable; there have been several impressive successes. Notably, image processing has seen rapid improvement and commercialization using similar techniques, as has the translation service that chess-playing computers failed to deliver on.[24] They still fall well short in terms of the strategic military decisions that were presaged for them, although progress continues.

Search algorithms like mathematical optimization (e.g., convex optimization) and metaheuristic search (e.g., genetic algorithm) can help us solve high-dimensional problems. For example, there is a similar effort using genetic algorithms to train fuzzy inference systems to control multiple agents in AFSIM.[25] RL takes a different approach, one that is less concerned with finding the "best" solution in a static world than with training agents to respond sensibly in dynamic environments. This makes RL better suited for "programming" agents that can interact with, shape, and be shaped by their environments: an autonomous car navigating a city block, for example. The advantage of a learning approach is that the rules of the game do not need to be explicitly or statically abstracted, represented and encoded. Instead, the agent can start with little or no knowledge about the game and simply learn to play by trial and error. Recent innovations, such as deep Q-learning—a key part of Google DeepMind's AlphaGo platform—leverages deep

[22] Volodymyr Mnih, Koray Kavukcuoglu, David Silver, Andrei A. Rusu, Joel Veness, Marc G. Bellemare, Alex Graves, Martin Riedmiller, Andreas K. Fidjeland, Georg Ostrovski, Stig Petersen, Charles Beattie, Amir Sadik, Ioannis Antonoglou, Helen King, Dharshan Kumaran, Daan Wierstra, Shane Legg, and Demis Hassabis, "Human-Level Control Through Deep Reinforcement Learning," *Nature,* Vol. 518, February 25, 2015.

[23] Dawn Chan, "The AI That Has Nothing to Learn from Humans," *The Atlantic*, October 20, 2017.

[24] Kaiming He, Xiangyu Zhang, Shaoqing Ren, and Jian Sun, "Deep Residual Learning for Image Recognition," *Proceedings of the 29th IEEE Conference on Computer Vision and Pattern Recognition: CVPR 2016*, 2016; Dzmitry Bahdanau, Kyunghyun Cho, and Yoshua Bengio, "Neural Machine Translation by Jointly Learning to Align and Translate," *arXiv*: 1409.0473, last revised May 19, 2016.

[25] Nicholas Ernest, David Carroll, Corey Schumacher, Matthew Clark, Kelly Cohen, and Gene Lee, "Genetic Fuzzy Based Artificial Intelligence for Unmanned Combat Aerial Vehicle Control in Simulated Air Combat Missions," *Journal of Defense Management*, Vol. 6, No. 1, March 2016.

neural nets to incrementally approximate the agent's reward functions and introduce hierarchical machine abstractions and compression into its reasoning.[26,27]

Some of the most impressive recent results have focused on games that are more closely related to military engagements. These systems can play over more continuous games with imperfect information and can account for issues logistics and short-term sacrifices to enable long-term goals. In the case of the video game Defense of the Ancients (DOTA) 2, these deep RL approaches have surpassed amateur humans and are approaching professional or superhuman levels. [28] In more-complex strategy games like StarCraft, the humans still have a substantial advantage at the time of writing, but many believe this advantage will not last much longer, given the history of successes in computerized gaming.[29]

These demonstrations in the gaming environment have been impressive, but as discussed earlier, the hope is that the techniques will find applications in the real world. Still, today, that aim has been a challenge and often an unfulfilled promise. The jump from a carefully prescribed environment to the messiness of the real world has been too great in most cases. With this project, we try to take an intermediate step in the transition from game environment to real-world utility. Simulated environments are widely used to explore new plans, designs, and tactics, and for training and as components in wargames. Simulations are, by their nature, constrained in much the same way that games are, but strive to incorporate enough of the real-world complexity to be useful to decisionmakers who consider the outcomes of the simulation.

By taking the deep RL algorithms and architectures that have recently been developed and demonstrated for such games as Atari, Go, and DOTA 2 and applying them to explore the Air Force standard simulation environment AFSIM, we hope to move these approaches a small intermediate step from games toward real-world utility. Although we recognize that simple air combat planning can be accomplished with a decision tree consisting of IF-THEN rules, we hope to explore the potential of RL tools to display unique behavior, generalize across different types of missions, and demonstrate multiagent cooperativity.

Organization of this Report

Chapter 2 discusses the tools and algorithms necessary to solve a simple, one-dimensional (1-D) mission planning problem. Chapter 3 follows a similar structure but moves the complexity

---

[26] Volodymyr Mnih, Koray Kavukcuoglu, David Silver, Alex Graves, Ioannis Antonoglou, Daan Wierstra, and Martin Riedmiller, "Playing Atari with Deep Reinforcement Learning," *arXiv*: 1312.5602, 2013.

[27] Arun Nair, Praveen Srinivasan, Sam Blackwell, Cagdas Alcicek, Rory Fearon, Alessandro De Maria, Vedavyas Panneershelvam, Mustafa Suleyman, Charles Beattie, Stig Petersen, Shane Legg, Volodymyr Mnih, Koray Kavukcuoglu, and David Silver, "Massively Parallel Methods for Deep Reinforcement Learning," *arXiv*: 1507:04296, last updated July 16, 2015.

[28] OpenAI, "OpenAI Five," blog post, *OpenAI Blog*, June 25, 2018.

[29] Yoochul Kim and Minhyung Lee, "Humans Are Still Better Than AI at Starcraft—For Now," blog post, *MIT Technology Review*, November 1, 2017.

forward into two-dimensional (2-D) space, which better represents real-world problems. Chapter 4 highlights some of the challenging computational infrastructure aspects of projects like this one. In Chapter 5, we summarize our findings and discuss potential future research directions. Appendixes provide detail on how various internal variables are handled by the tools used in this effort.

A large aspect of this work was exploratory, and we pursued several lines of effort. Figure 1.2 lays out a roadmap of these efforts.

Figure 1.2. Roadmap of the Report

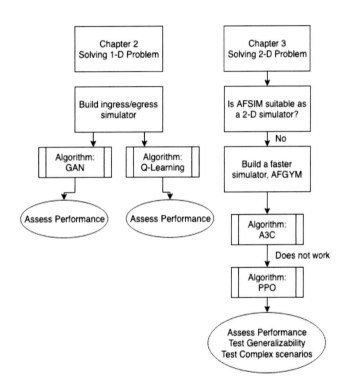

NOTE: A3C = Asynchronous Advantage Actor Critic; PPO = proximal policy optimization. AFGYM is the authors' name for a fast, low-fidelity version of AFSIM.

2. One-Dimensional Problem

ML projects can rapidly grow in complexity. It can be easy to convince oneself that a program works when it does not or to create a complicated program that fails in nonintuitive or interpretable ways. To minimize those risks, we took an incremental approach, starting with a highly simplified mission-planning problem and building up from there toward increasing levels of complexity. This simplified approach started with a 1-D navigation problem and well-established, theoretically sound algorithms that have a long history of empirical validation. Because the optimum solution to this 1-D problem can easily be found by hand, it also allows us to easily verify whether an algorithm is performing correctly.

Problem Formulation

The simplified 1-D scenario has three components: a fighter aircraft , a jammer aircraft, and a SAM system (Figure 2.1).[30] The objective of the fighter is to destroy the SAM. The objective of the SAM is to destroy the fighter. In this formulation, the SAM holds the advantage in range (100 km) and therefore can destroy the fighter (80 km) before the fighter can fire on the SAM. The case shown in the figure has only two potential outcomes. The fighter can be sent out some distance and turn back before reaching the SAM's range, so both survive, or the fighter can be sent into the SAM's range and be shot down.

Figure 2.1. One-Dimensional Problem Without Jammer

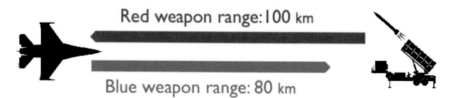

Red weapon range: 100 km

Blue weapon range: 80 km

However, timing and placement are important. Adding a jamming aircraft can return the advantage to the fighter by increasing the number of possible outcomes (Figure 2.2). The jammer can approach the SAM and reduce the SAM's range. If the fighter enters during that period, it can destroy the SAM and live. The jammer can also fly too close to the SAM before the fighter destroys it, or the jammer, the fighter, or both could be shot down. The AI system is given the location of the SAM and the range of all three components (fighter, jammer, and SAM) and is

---

[30] We model the aircraft as similar to an F-16 fighter jet, which often performs the counter-SAM mission for the U.S. Air Force.

expected to learn what times and distances to send out the fighter and jammer to successfully destroy the SAM without losing either aircraft.[31]

Figure 2.2. One-Dimensional Problem with Jammer

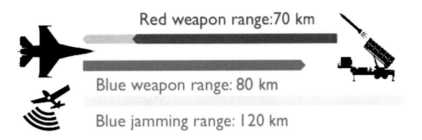

General mission-planning in the AFSIM environment would require many degrees of freedom in plan specification, but for this 1-D case, mission specifications were restricted to motion- and action-planning in one effective dimension of movement. The mission scenarios or laydowns are embedded in the standard three-dimensional (3-D) environment, but the control variables specify action and motion in only one dimension. Table 2.1 details the agents in the 1-D AFSIM scenario.

Table 2.1. The 1-D Mission Scenario Agents

Side	Agent Description
Blue	• Mobile fighter aircraft with missile-attack capability within a specified firing range
	• Mobile electronic jammer aircraft capable of nullifying or attenuating Red SAM's area of attack
Red	• Stationary SAM battery with a predefined (disclosed to Blue) area of attack

The scenario is 1-D because all agents are constrained to fly on the same line connecting the Blue and Red starting locations. Altitude is held fixed for the Blue aircraft. To ensure that we provide the AI system with sufficiently diverse experiences from which to generalize, the Red location is initially offset from the fighter's starting position by a randomly chosen distance. The AI system that learns to solve this scenario would need to learn the following information:

1. Send out the jammer before the fighter to disable the Red SAM.
2. Hold the jammer at a distance from the SAM that is close enough to be effective but not so close as to get shot down.
3. Send out the fighter to attack the disabled SAM without getting too close.
4. Return both aircraft to base.

Figure 2.3 shows the 1-D scenario rendered in the AFSIM environment from a view above and to the side of the three entities. The Blue aircraft are moving from left to right here. The Blue

---

[31] For our initial efforts, the AI planner was provided the exact location of the SAM and Blue aircraft, although this information was not required.

fighter trails the jammer that is successfully jamming the Red SAM (this action can be seen by the orange jamming line and the fact that the yellow detection lines flow only from the Blue platforms toward the SAM). This particular example goes on to show the fighter successfully flying in, shooting the SAM, and turning around.

Figure 2.3. Screenshot from VESPA Playback of a 1-D Scenario Example

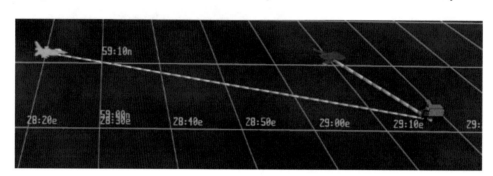

Each scenario is defined by two sets of variables: environmental and learning. The first set includes features of the scenario that are predetermined. Setting the variables' values describes a specific laydown, or initial "state of the world." For example, the firing range on a platform is an environmental variable: Setting its value helps define the environment in which the ML agent is acting. Learning variables are features that the ML agent learns. Setting their values describes a specific plan or way by which platforms will interact with the environment. For example, the distance a platform flies can be a learning variable and might depend on features in the environment, such as firing ranges of nearby platforms. In the 1-D mission-planning problem, the ML agent seeks the optimal plan for a given laydown. Platform behaviors and scenario variables are described in Tables 2.2 and 2.3.[32] The goal of varying firing ranges and distances is to teach the fighter a generalizable strategy (e.g., ingress if fighter range is greater than that of the SAM's; egress otherwise) rather than learn a hard-coded strategy specific to one scenario.

---

[32] Note that although the fighter and jammer start at exactly the same the location, we are not modeling any collisions.

14

Table 2.2. 1-D Problem Formulation—Environmental Variables

Environmental Variable	Variable Bounds
Fighter firing range	0–40 km
Fighter speed (fixed)	740 km/hour
Jammer range (fixed)	30 km
Jammer speed (fixed)	740 km/hour
SAM firing range	0–40 km
SAM distance from fighter (offset)	0–100 km
Fighter starting position (fixed)	Latitude: 59.04° Longitude: 27.75° Altitude: 6 km
Jammer starting position (fixed)	Latitude: 59.04° Longitude: 27.75° Altitude: 6 km

Table 2.3. 1-D Problem Formulation—Scenario Set Up and Learning Variables

Agent	Side	Learning Variables	Scripted AFSIM Behavior	Behavior to Learn
Fighter	Blue	• Ingress distance • Lead distance for jammer to fly before fighter starts ingress	• Fighter waits until jammer flies lead distance before it begins its ingress • If the SAM is within the fighter's firing range when it reached its ingress distance, fighter fires at SAM, then turns around • If the SAM is not in fighter's firing range when it hits its ingress distance, fighter turns around and heads to starting position (if not already shot at)	• Learn to fly no further than its firing range in any scenario • Learn when jammer will be effective in ensuring safe passage toward the SAM, even when SAM initial firing range is greater than the fighter's
Jammer	Blue	• Ingress distance	• Jammer starts ingress toward SAM at scenario start • If jammer gets within 30 km of SAM, jammer deploys s-band jamming • Jammer stops and hovers after reaching ingress distance	• Learn to fly up until its effective range and no further
SAM	Red	• None	• If Blue platform gets within SAM's firing range, SAM shoots at platform	• None

The training data for the AI system's adversarial learning consisted of about 20,000 randomly generated laydown and plan pairs. Plans were randomly assigned to the laydowns so that the agent could initially assess which plans proved successful against their environment. Therefore, for the training data, both environmental and learning values were provided. The environmental variable values were randomly chosen within the bounds given in Table 2.2. Given a laydown, its plan values (or what were later learned by the ML agent) were randomly generated within the bounds given in Table 2.4 below.

Table 2.4. 1-D Problem Formulation—Learning Variable Bounds for Test Data

Learning Variable	Variable Bounds
Fighter Ingress Distance	Minimum: 0 km
	Maximum: SAM location (this upper bound is set so that the fighter cannot fly past the SAM)
Jammer Ingress Distance	Minimum: 0 km
	Maximum: SAM location (this upper bound is set so the jammer cannot fly past the SAM)
Jammer Lead Distance	Minimum: 0 km
	Maximum: Jammer ingress distance (the lead distance—or distance the fighter waits for jammer to fly before it begins its ingress—cannot exceed the total distance the jammer ingresses)

Machine Learning Approaches for Solving the 1-D Problem

We can specify the learning problem as follows: *Given a mission scenario, learn to produce a mission plan that is judged successful via the AFSIM mission simulation.* A number of viable ML modeling approaches can be used to tackle this problem. More generally, many search methods could also be used to solve the decision variables using iterative or batch evaluations, a fact that highlights the overlap between search and ML in simple planning problems.

We chose to formulate the problem as an imitative learning problem using the AFSIM simulator as the source of data on simulated plans. The basic imitative learning model ingests positive exemplar data (i.e., successful missions), then learns a latent space representation and generator for similar positive examples. We adapted that learning model with a state input vector to create a state-conditional plan generator. This characteristic is important because we want our plan generator to be flexible and generalized enough to solve for input scenarios (primarily different starting locations and engagement ranges in this simple case) that it has never encountered.

Two algorithms were applied at this stage because it was not clear a priori which would be the most suitable. Both have been very well studied theoretically and have long histories of application. The first, Q-learning, is the approach that was used in beating Atari games.[33] Here, the neural network was a simple fully connected feed-forward network with a few layers and a few hundred nodes total.

In this 1-D example, a decision must be made at the start that depends only on the ranges of the fighter, jammer, and SAM, and the location of the SAM. Therefore, the *state* is the positions of the two aircraft and the SAM. The *action* is the specification of the timing and turning points of the fighter and jammer, and the *value* is the score that is received once the run is over. To encode this problem into machine code, this problem was reformulated as a step-by-step process during the flight, in which the fighter and jammer decide after each step whether to proceed

---

[33] Mnih et al., 2013.

16

forward, stay in place, or turn back. In that case, the state would be the locations of the fighter, jammer, and SAM, the action would be to proceed forward, stay, or turn back, and the value would be a number describing whether the current state is likely to achieve a favorable outcome. By framing the decisionmaking process into a step-by-step approach, Q-learning can be applied.

In Q-learning, the neural network is used to approximate what is called the action-value function. This method contrasts with other methods that estimate the value of being in a given state and "on-policy" methods, such as state–action–reward–state–action.[34] Put simply, Q-learning trains a neural network that calculates the value that will result from a given action, and that score can be compared across all of the available actions to select the most beneficial option.[35] The goal of Q-learning is to learn and approximate this Q-function for all possible states and actions.

The Q-learning approach is fairly robust, and for a problem as simple as this 1-D one, the given network has a high degree of capacity for learning. Once the hyperparameters, such as learning rate, batch size, and exploration parameters were set, the network was able to improve rapidly, as shown in Figure 2.3, where the error drops immediately and afterward decays rapidly over just a few hundred batches. Error is defined by how close the neural network is to the true Q-function and can be interpreted as a proxy for how well the RL is performing (lower is better).

Figure 2.4. Q-Learning Proceeds Rapidly on One-Dimensional Problem

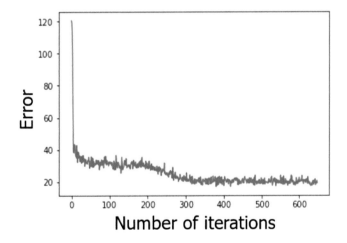

Q-learning showed promise, as expected, although the simplicity of this initial problem somewhat limits its utility. With only a single step per episode (namely, selecting mission parameters), other approaches were considered more appropriate for continued investigation

---

[34] Richard S. Sutton and Andrew G. Barto, "Chapter 6: Temporal-Difference Learning," *Reinforcement Learning: An Introduction*, 2nd ed., Cambridge, Mass.: MIT Press, 2017.

[35] Refer to Appendix D for an in-depth explanation and mathematical formulations.

rather than optimizing performance of the Q-learning algorithm once initial success had been demonstrated. One of these other methods used GAN.[36]

The second ML approach implemented was a GAN model that trains a generator and a discriminator (or critic) network to solve an imitation-learning problem. The generator subnetwork takes in noise input and puts out imitation candidates (in this case, mission plans). The discriminator subnetwork takes the imitation candidates as input and puts out a continuous rating on a [0,1] interval that indicates the quality of the imitation. Both subnetworks are updated via backpropagation on discriminator error signals in response to positive exemplar data. The state-conditional plan generator is simply an adapted implementation of a conditional GAN (CGAN).[37] Figure 2.5 shows the architecture of our adapted state-conditional GAN model, which implements a solution for a one-step MDP. This perspective is consistent with the recent work identifying connections between GAN models and actor-critic reinforcement learning models (e.g., in the 2-D planners developed for this report) for solving multistep MDPs.[38]

The CGAN can be interpreted as two networks competing against each other. One network (denoted by G) is designed to generate "fake" plans given a scenario (fake is defined as non–human generated). The other network, denoted by D, is designed to discriminate between real and fake plans conditioned on a scenario[39].

Agent-environment behavior was reformulated to accommodate the CGAN. Unlike the step-by-step breakdown for Q-learning, the action is a single input, ingress distance. The generator, G, draws this ingress value, z, from a probability distribution conditioned on the scenario, y. The 1-D environment interprets this value as ingress the appropriate distance, attempt to fire, and then egress.

---

[36] Ian Goodfellow, Jean Pouget-Abadie, Mehdi Mirza, Bing Xu, David Warde-Farley, Sherjil Ozair, Aaron Courville, and Yoshua Bengio, "Generative Adversarial Nets," in Z. Ghahramai, M. Welling, C. Cortes, N. D. Lawrence, and K. Q. Weinberger, eds., *Advances in Neural Information Processing Systems (NIPS 2014)*, Vol. 27, 2014.

[37] Mehdi Mirza and Simon Osindero, "Conditional Generative Adversarial Nets," *arXiv*:1411.1784, 2014.

[38] David Pfau and Oriol Vinyals, "Connecting Generative Adversarial Networks and Actor-Critic Methods," *arXiv*:1610.01945, last updated January 18, 2017.

[39] Refer to Appendix D for an in-depth explanation and mathematical formulations.

Figure 2.5. Imitative Planning Model Architecture to Solve 1-D Problem

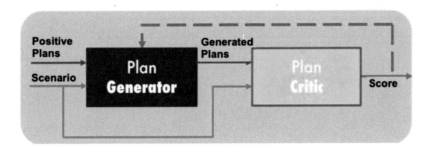

Evaluative Feedback for GAN Planner Training

Simulation training data consist of plan vectors with randomly assigned laydowns or scenarios. The laydown and the plan combine to specify a *mission vector*. Training the planner requires evaluative feedback from the AFSIM simulator on the mission outcome (success or failure of the specified mission). Because the interface between the AI system and AFSIM was somewhat unwieldy and mission success in our simple example can be calculated directly, we developed a simpler and faster test environment. This simple 1-D mission simulator was implemented in Python and called AFGYM.[40] We use it here to estimate the AFSIM outcome of a mission vector and also score mission outcomes. This allowed us to remove AFSIM from the algorithm development path, other than for generating starting positions. The scoring function rewards survival of Blue agents (fighter and jammer at +2 and +1 reward points, respectively) and penalizes the survival of Red agents (terms of reference at −4 reward points). We use this evaluator to help select the randomly generated simulation mission for successful mission vectors. Recall that GAN training requires only positive exemplar data for training. The positive AFSIM exemplar training data for our GAN mission planner consisted of about 26,000 successful randomly generated laydown and plan pairs.

This ad-hoc solution to the interface problem incurs a complexity cost. The initial data subselection implies that our GAN planner is actually learning to solve the planning problem for the simplified evaluator. The subsequent success of the GAN planner's solutions in the AFSIM environment is really a demonstration of successful *transfer learning*.

Evaluating the GAN Planner

Evaluating the quality of the GAN planner requires the formulation of relevant figures of merit for planning models. We use the concept of "lift" to quantify the effects of using the GAN against an agent acting at random. Table 2.5 identifies some of our relevant metrics (these metrics are not exhaustive).

---

[40] This is a play on the words "AFSIM" and the "Gyms" commonly used as toolkits for developing ML algorithms. AFGYM is described in more detail in the next chapter.

Table 2.5. Description of Evaluation Metrics for ML Planning Models

Metric	Description	Baseline (percent)
Lift over failed samples	What percentage of the randomly generated mission vectors that originally scored as failed does the trained planner make successful?	0
Lift over positive samples	What percentage of the randomly generated mission vectors that originally scored as successful remains successful after we apply the trained planner?	100
Overall lift	What percentage of the total randomly generated mission scenarios does the trained planner make successful? (Baseline is defined by the success rate of a purely random agent.)	36
Training set size	How many samples did the planner train on to achieve the reported performance numbers?	n/a

To test the GAN architecture and accelerate development, we initially benchmarked this approach on AFGYM. After development, we then switched back to the AFSIM environment to score mission-planning performance. The first column of Table 2.6 reports the performance metrics of the simplified AFGYM model. The second column of numbers reports the metrics for the 1-D AFSIM GAN planner.

Table 2.6 shows that the GAN planner is learning better planning behavior than would a random untrained planner. It produces mission plans with scores higher than a random planner. However, higher scores do not always translate to better observed behaviors, even though the reward metric is designed to encourage Blue survival and strongly discourage Red survival. Further inspection of the survival statistics shows that the GAN planner more likely improved its understanding of when a scenario is difficult or impossible to solve. The planner then devised conservative solutions for these scenarios, where Blue does not try to strike the SAM.

Table 2.6. Performance of GAN Planning Models on Two Mission Environments

Metrics	Simplified Model	AFSIM
Lift (over failed samples)	+30%	+17.9%
Lift (over positive samples)	−2.1%	−6.8%
Lift (overall)	+26%	+11%
Training set size	50,000	~20,000

3. Two-Dimensional Problem

The 1-D problem in the previous section represents a proof-of-concept demonstration of the potential of GANs in simple planning problems. The next step is to extend the problem into two dimensions, which creates the need to route-plan from a starting location to a target engagement location, avoiding threats along the way. Essentially the go-or-stop problem of the 1-D problem is now a time series of "how much should I turn" questions. There are different ways to represent the routing problem for search and learning, ranging from graphs over discrete grids to parameterized continuous representations like Bezier splines. Here, we leverage the commonly deployed approach of formulating the routing problem as an agent-based MDP, where the UAVs take in the collective state of the world at each time step and use this input to formulate their next actions. More formally, this is known as a partially observable Markov decision process (POMDP) in which agents do not have perfect knowledge of the actual state of the world.[41] In this initial formulation, the agent controls all of the UAVs and has access to perfect information. This incremental and iterative approach stands in fundamental contrast to the one-shot planning strategy in the 1-D case. Here, the choice of the MDP formulation is driven by the need to simultaneously manage the size of the state space and maintaining granularity in the route representation. Moreover, the MDP formulation can also more efficiently support real-time planning to dynamic events.

Problem Formulation

The general 2-D SEAD problem is more complex than its 1-D counterpart because of the vastly larger decision space, and, depending on the agents present, can easily be too complex to "solve" by inspection or even significant computational resources. We further add to the complexity of the problem by introducing a Red target (in the 1-D problem, the SAM was considered the target) that can be attacked by Blue weapons by flying around SAMs.[42] The problem formulation and decision space are illustrated in Figure 3.1. Here, a Blue UAV starting from some random location in 2-D space has to learn how to strike the Red target while coping with the presence of randomly placed SAMs. The Blue agent here has perfect information about

---

[41] For more detail, see K. J. Åström, "Optimal Control of Markov Processes with Incomplete State Information," *Journal of Mathematical Analysis and Applications*, Vol. 10, No. 1, February 1965. For a recent application, see Haoyu Bai, David Hsu, Mykel J. Kochenderfer, and Wee Sun Lee, "Unmanned Aircraft Collision Avoidance Using Continuous-State POMDPs," in Hugh Durrant-Whyte, Nicholas Roy, and Pieter Abbeel, eds., *Robotics: Science and Systems VII*, Cambridge, Mass.: MIT Press, 2012

[42] For obvious reasons, "flying around" was not an option in the 1-D case.

the locations of the Red missile and targets but no scripted rules and heuristics for striking the target while avoiding the SAM. The goal is to learn the listed heuristics through repeated play.

In AFSIM form, the basic 2-D problem has three platforms: a Blue fighter, a Red SAM, and a Red target. The mission objective is for the fighter to reach and shoot the target while avoiding being shot down by the SAM. Figure 3.1 shows an example 2-D laydown. The fighter is in the lower left, the SAM is in the center, and the target is in the top right. The mission objective is for the fighter to fly around the SAM's firing range and hit the target. In this simplified form, with only two agents and a target, we can easily verify algorithm performance.

Figure 3.1. Example Screenshot of the 2-D Laydown

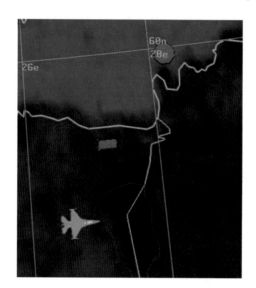

As opposed to the 1-D problem, which used RL and GAN methods to choose a full plan for a given laydown, the 2-D problem uses RL to inform platform actions throughout the scenario. This process involves ongoing coordination between AFSIM and the ML agent during mission execution itself. AFSIM reports the scenario state at intervals throughout the run, the agent returns actions for AFSIM to carry out in the following interval, and the process repeats until the simulation completes. This coordination is described in more detail below.

The environmental variables (or those that define the initial "state of the world" for a given run) for this 2-D scenario are listed in Table 3.1.

22

Table 3.1. 2-D Problem Formulation—Environmental Variables

Environmental Variable	Variable Bounds
Fighter location	Latitude: 58.5° to 59.0363° (random)
	Longitude: 26.3° to 26.8° (random)
SAM location	Latitude: 59.1° to 59.5° (random)
	Longitude: 27.2° to 27.5° (random)
Target location	Latitude: 59.6° to 60° (random)
	Longitude: 27.7° to 28.5° (random)
Fighter firing range	3.5–10.5 (in 100km x 100km game area)
SAM Firing Range	5.0–10.5 (in 100km x 100km game area)

The action given out by the agent contains values for the learning variables. In this problem, the only learning variable—or the only value delivered by the agent at each time step—is the fighter's next heading. The fighter will proceed to fly in a straight line in this heading direction until the next state readout is passed to the agent. It should be emphasized that although there was one ideal value per learning variable (and laydown) in the 1-D case, here the agent is learning the ideal fighter heading at each time step based solely on the current scenario state and nothing prior.

Reduced Order Simulation Environment: AFGYM

Parameterizing 2-D motion requires continuous variables (aircraft heading, in degrees). For simplicity, velocity was kept constant to limit the size of the decision space. The introduction of a continuous heading variable meant that an ML process needed substantially more exploration than the 1-D case to learn how to navigate a laydown. The high number of simulations required to conduct such an exploration prompted the creation of a faster "AFSIM-light" program. This approach is commonly done in the field. A notable example is the OpenAI DOTA 2 bot, which was trained using sped-up simulations to achieve 180 years of gameplay per day.[43] The 2-D model addressed the model-simulator interface problem by expanding the 1-D reduced-order training simulation, AFGYM, into two dimensions. A key research question is then the transferability of the resulting agent from the proxy environment (AFGYM) to the representative one (AFSIM).

AFGYM is a Python module created to facilitate the 2-D deep RL problem. AFGYM was designed to simulate the bare minimum to demonstrate SEAD behavior (simulate aircraft movement around enemy SAMs). AFSIM was designed as an all-in-one air-combat simulator and performs many extraneous calculations with respect to the SEAD mission. AFGYM performs a major simplification; it does not perform any radar calculations. Both Blue and Red agents automatically fire if any valid target is within a predefined weapon range. EW also is not modeled, and jamming is calculated as a percent reduction of weapon range. The AFGYM application programming interface conforms to OpenAI's Gym environment, which is a

---

[43] James Vincent, "AI Bots Trained for 180 Years a Day to Beat Humans at Dota 2," *The Verge*, June 25, 2018.

standardized Python toolkit and target environment designed to facilitate and accelerate RL research.[44] OpenAI Gym uses a set of standardized protocols to allow a researcher to prototype AI systems that can accomplish such tasks as balancing an inverted pendulum, playing Atari games, and controlling a robot arm. AFGYM is a module that uses these same protocols to simulate simplified AFSIM behaviors. AFGYM supports a variety of RL approaches, such as parallelized learning, and allows for fast testing of algorithms.

Specifically, AFGYM currently supports the creation of Blue aircraft, for which users define missile ranges, missile count, missile cooldown time, jamming ranges, and jamming effects. Red adversaries are stationary platforms that allow for user-defined missile ranges, missile count, and missile cooldown time. These adversaries consist of stationary SAMs and targets. A notable simplification from AFSIM is that AFGYM does not perform floating-point radar-detection calculations, and instead simply uses table lookup detection ranges for sensors. Missile and jamming ranges are assumed to be constant, and, as with AFSIM, weapons do not fail to remove unwanted uncertainty. AFSIM has a detailed set of programming rules available to control agent behavior that AFGYM does not attempt to duplicate, again for speed.[45] Furthermore, AFGYM radar ranges are assumed to be omnidirectional, whereas AFSIM factors in radar orientation, scan rate, and antenna pattern. All numbers within AFGYM are unitless and correspond to the Cartesian coordinate system (e.g., a missile range of 10 corresponds to 10 units on the coordinate system) on a flat plane (i.e., not a spherical Earth as AFSIM uses). Units might freely traverse a square-sized sandbox within a range of [0,100]. At each timestep, AFGYM reports the status of each unit (e.g., location, heading, weapon ranges), and that information is then fed into a neural network. To prevent the common issue of exploding gradients during training, we normalized all AFGYM values to [0,1] for training. Unnormalized multiscaled inputs tend to cause large gradients that saturate network layers at the backpropagation step.

Machine Learning Approaches for Solving the 2-D Problem

We shifted toward policy gradient–based algorithms to solve the 2-D problem. For complex models, as defined by large state and action spaces, policy-gradient algorithms outperform off-policy algorithms, such as Q-learning. Policy-gradient methods have convergence guarantees and learn high-dimensional spaces better than do off-policy methods.[46] Q-learning is typically not applied on problems with continuous action spaces for these reasons. This was seen in practice because we tested the REINFORCE Q-learning algorithm on continuous 2-D problem formulation and the resulting behavior was a random walk. Additionally, GANs are unsuitable

---

[44] As of September 2018, many of these gyms are hosted.

[45] For Blue behavior, the agent under development serves as the replacement for this behavioral logic.

[46] Kai Arulkumaran, Marc Peter Deisenroth, Miles Brundage, and Anil Anthony Bharath, "A Brief Survey of Deep Reinforcement Learning," *IEEE Signal Processing Magazine*, Vol. 34, No. 6, November 2017.

for the 2-D formulation because decisions are made in "real time" (at each step) rather than planning the mission fully a priori.

Asynchronous Advantage Actor Critic

The first RL algorithm that demonstrated promise was A3C,[47] which builds on classic actor-critic models for RL with innovations that allow for improved training parallelization and better management and utilization of training experiences. When this method was discovered, researchers showed that it outperformed many other existing algorithms in terms of performance (such as game score) and learning efficiency. A3C offers several advantages over traditional methods that directly use rewards.

Briefly, A3C builds on the REINFORCE algorithm, which was designed as an on-policy algorithm. In contrast to Q-learning, which effectively amounts to the neural network storing a large table of state-action pairs, on-policy algorithms use a neural network to learn both the Q-function and the policy directly.[48] A3C uses a computed value called *advantage*, which is defined as the difference between the machine's expected reward for taking an action and the actual reward received. By using advantage, the algorithm uses relative rewards to update itself. For example, a tabula rasa algorithm learning the Pong video game will move a paddle randomly left and right. Its initial value network will calculate any action to have a value of zero because of past experiences. However, the algorithm might randomly move toward the ball, bounce it back, and receive a positive reward that was an improvement over its expected reward of zero. Using relative rewards (compared with past performance) rather than absolute rewards helps the algorithm efficiently learn better strategies. This network is called the critic.

Second, A3C maintains a separate neural network, called an actor network, to compute the best action to take at each step. After an action is taken, the critic network calculates the advantage. The advantage, in conjunction with an entropy term to encourage exploration, is used to conduct backpropagation into the policy network to increase the likelihood of actions that yield positive advantages and to decrease the likelihood of actions that yield negative advantages.

Finally, A3C's innovation is to parallelize the learning process via asynchronous updates. Instead of waiting for a single central processing unit (CPU) to serially run simulations, calculate advantages, and update the actor-critic network, A3C leverages multiple CPUs for faster learning. Each worker CPU iteration consists of downloading a copy of the most recent global network, running simulations, and performing asynchronous learning updates to the global network.

---

[47] Volodymyr Mnih, Adrià Puigdomènech Badia, Mehdi Mirza, Alex Graves, Timothy P. Lillicrap, Tim Harley, David Silver, and Koray Kavukcuoglu, "Asynchronous Methods for Deep Reinforcement Learning," in Maria Florina Balcan and Kilian Q. Weinberger, eds., *Proceedings of the 33rd International Conference on Machine Learning*, Vol. 48, 2016.

[48] Refer to Appendix D for an in-depth explanation and mathematical formulations.

As discussed later in the results section, A3C showed considerable promise when training various mission-planning laydowns in AFGYM. For example, A3C could succeed in missions where it had to control multiple Blue planes (multiple fighters, fighter and decoy, and multiple fighters + jammer) against multiple Red adversaries.[49] Trained algorithms exhibited synergistic effects, such as: (1) jammer reduces the range of a SAM and allows an otherwise outranged fighter to get close and eliminate the SAM, and (2) multiple fighters with differing missile ranges engages the appropriate Red targets. However, there were issues with consistency (not all simulations for a laydown succeed) and with convergence. *Policy collapse* is a phenomenon wherein the actor-critic policy network diverges from an optimal policy and stops learning from experience. Neural networks serve as function approximators to the policy function, π. Policy collapse occurs when the neural network diverges away from the policy function because of numerical issues during the backpropogation step.[50] This was observed when most A3C runs converged on the behavior of planes spinning in stationary circles.

The tabula rasa A3C algorithm controls the Blue aircraft via random walk. The algorithm must take a long series of correct actions to avoid SAMs and eliminate a target to experience a positive reward. Many of the random walks that position a Blue plane in the general vicinity of a Red SAM result in a Blue plane getting shot down by the SAM. As a result, the algorithm penalizes exploration into that area and moves aircraft away in subsequent simulations. Attempts to penalize this type of behavior via the reward function resulted in Blue planes spinning in stationary circles (this behavior attempts to simultaneously avoid a SAM and avoid the "flying away" penalty). Given the rarity of exploring near a SAM, policy updates that result from this exploration are outweighed by the other asynchronous workers discovering the local optima of spinning in circles. This issue was addressed via curriculum learning, where laydowns become progressively difficult. This approach taught the algorithm how to achieve rewards (planes that have superior missile range fly toward SAMs; otherwise avoid SAMs and seek targets only) by initializing aircraft close to Red targets and SAMs. As training iterations progress, laydowns initialize Blue planes further from their targets and the algorithm adapts accordingly.

Trust Region Policy Optimization and PPO

Trust region policy optimization (TRPO) and PPO improve on gradient-based methods of RL. Gradient-based methods traditionally seek to optimize a cost function (say, maximize rewards or advantage) via gradient descent. This technique is prone to slow learning or to policy (learning) collapse, depending on the user-defined learning rate. TRPO addresses this issue by

---

[49] Decoys were modeled as identical to the fighter aircraft, but their loss was not counted as a failure of the mission. Their purpose is to distract the Red SAM and provide a small window of vulnerability that Blue aircraft could take advantage of if the timing and geometry were correct.

[50] Exploding or vanishing gradients typically cause this issue. Backpropogation steps rely on estimated gradients. Should the gradient be too large or small, the neural network parameters are affected and diverge away from the desired policy function.

maximizing the same objective function, J, as that of the A3C case, but constrains the optimization such that the policy, π_θ, does not change beyond a certain threshold after the update step. This is accomplished by limiting the Kullback-Leibler divergence (KL) between the pre-update π_θ and the post-update $\pi_{\theta'}$.[51] This constrained optimization is intractable to calculate in practice and the TRPO method represents one such estimation using a surrogate objective function. We refer the reader to the original paper for an in-depth explanation of TRPO.[52]

This constraint helps prevent learning collapse, where the effects of a bad update (such as reinforcement of circle-spinning behavior) are limited. Furthermore, the surrogate objective function is a quadratic approximation of the cost function and guarantees improvement. PPO was developed by OpenAI as a simplified way to implement TRPO because of the complexities of calculating the KL statistic.[53] The innovation made by PPO was to eliminate the calculation of KL entirely by clipping the advantage term within the objective function. Clipping has the effect of tempering an update step regardless of whether the advantage term was extremely high or low. In A3C, such extreme behavior results in exploding or vanishing gradients, respectively, and causes learning collapse.

A nonparallelized version of the actor-critic network was developed for use with the PPO algorithm. PPO was used to update the actor network, and regular gradient descent was used for the critic network.

Agent and AFSIM Coordination

Once agents have been developed with AFGYM in the MDP formulation of the problem, AFSIM is executed in a stepwise fashion to update the state of the world.[54] The ML agent then consumes the state vector and generates actions for the next step. This execution loop continues for the duration of the run or until the fighter is shot down. The elements of the state vector are summarized in Table 3.2.

---

[51] John Schulman, Sergey Levine, Philipp Moritz, Michael Jordan, and Pieter Abbeel, "Trust Region Policy Optimization," in Francis Bach and David Biel, eds., *Proceedings of Machine Learning Research*, Vol. 37: *International Conference on Machine Learning*, 2015; and John Schulman, Filip Wolski, Prafulla Dhariwal, Alec Radford, and Oleg Klimov, "Proximal Policy Optimization Algorithms," *arXiv*: 1707.06347, last updated August 28, 2017.

[52] Schulman et al., 2015.

[53] Schulman et al., 2017.

[54] This does not mean the AFSIM scenario "stops" and "starts." Because AFSIM carries out events linearly, all that is needed for AFSIM to report its current state to the agent is a predefined event trigger (in our case, the fighter flying a set distance in normalized space). AFSIM's next event—reading in a new fighter heading and turning to that direction—will not occur until the agent's action is delivered, as this is set as the next "event" in the AFSIM run.

Table 3.2. 2-D Problem Formulation—State Reported to ML Agent at Each Time Step

State Variable	Explanation
Fighter heading	Direction that fighter is pointing (0' is the positive x direction in Cartesian coordinates); -180 to 180 degrees, normalized to be in [0,1]
Fighter latitude	Latitude of fighter (y value in Cartesian coordinates); normalized to be in [0,1]
Fighter longitude	Longitude of fighter (x value in Cartesian coordinates); normalized to be in [0,1]
Fighter dead	1 if dead, 0 if alive.
Fighter firing range	Maximum striking (Euclidean) distance to a target; normalized to be in [0,1].
Fighter weapon count	Ammunition count; normalized to be in [0,1].
SAM pointing	Not currently used (implemented for future use when radar could be pointed toward a heading); 0—fixed report to agent.[a]
SAM latitude	Latitude of SAM (y value in Cartesian coordinates); normalized to be in [0,1]
SAM longitude	Latitude of fighter (x value in Cartesian coordinates); normalized to be in [0,1].
SAM dead	1 if dead, 0 if alive.
SAM firing range	Maximum striking (Euclidean) distance to a target; normalized to be in [0,1].
SAM weapon count	Ammunition count; normalized to be in [0,1].
Target heading	Not currently used (implemented for future use when radar could be pointed toward a heading); 0—fixed report to agent.
Target latitude	Latitude of target (y value in Cartesian coordinates); normalized to be in [0,1].
Target longitude	Latitude of target (x value in Cartesian coordinates); normalized to be in [0,1].
Target dead	1 if dead, 0 if alive.

[a] AFGYM simplified SAM behavior and did not consider SAM pointing directions. SAMs were assumed to have omnidirectional detection behavior, and the neural network was not trained to interpret SAMs as having a pointing direction.

NOTE: All values in the Current State vector are normalized to be between 0 and 1 because it bounds the backpropogation gradients during neural network training . See Appendix A for normalization details.

The successful implementation of the interaction between the agent and simulator relies to a great extent on the flexibility and state transparency of the AFSIM environment. The implementation of the MDP formulation presents a generalizable and interesting case study on how to integrate ML agents with existing simulation environments. Appendix C provides more details on the coordination and message passing implementation.

2-D Problem Results

After training, by matching AFSIM and AFGYM's output formats at each time step, we successfully used the pretrained agent in coordination with the AFSIM environment (thereby removing AFGYM from the process). Algorithms trained via A3C, and on simple laydowns (such as a single fighter and a SAM) suffered from overfitting. These algorithms were vulnerable to failure when minor adjustments were made to the laydowns (such as a slight change in initial fighter heading). A3C algorithms could learn some successful behavior in multiagent laydowns (multiple fighters or fighter and jammer or decoy), but movement was erratic, and success was rare (occurred in less than 10 percent of the cases). Furthermore, all algorithms eventually suffered learning collapse or convergence to local minima. Tapering the learning rate helped somewhat, but did not prevent this degradation.

The largest issue with A3C was the lack of generalizability. Trained algorithms were unable to demonstrate consistency in winning their trained laydown scenario and they adapted poorly to new scenarios. As seen in Figure 3.3, trained algorithms did not respond well to minor perturbations in the laydown, such as different Red target positions or initial aircraft headings. The result was erratic flight behavior that failed in these laydowns (flying away from Red targets or getting shot down).

Figure 3.2 shows four AFGYM simulations that highlight the situations described above. An AFGYM simulation occurs on a 100km x 100km grid, and blue lines represent a fighter's trajectory. The red diamond with an associated red circle represents a SAM and its effective range. In panel (A), there are multiple fighters available, but only one has sufficient missile range to safely eliminate the SAM. Despite the winding paths, the algorithm selects the correct fighter to ingress. In this snapshot, the fighter closest to the SAM is about to successfully attack. Panel (B) shows a rare success case of a fighter and jammer laydown. In this scenario, all of the fighters are outranged by the SAM and can only eliminate the SAM if the jammer is present. We see a jammer (green trajectory) circling a safe region near the SAM and reducing the SAM's effective range. In this snapshot, the closest fighter ingresses and is about to eliminate the SAM. Panel (C) demonstrates a typical fighter and jammer laydown. Like panel (B), this laydown is designed such that the fighter can eliminate the SAM only in the presence of a jammer. The A3C algorithm struggled to learn timing. We see two instances in which the jammer is close enough to reduce the SAM's range and the fighter ingresses. Unfortunately, the jammer flies too far (or is shot down) and the fighter maneuvers away in both cases. Panel (D) visualizes learning collapse. Saturation of certain connections in the neural network causes the algorithm to output a single value telling the planes to turn counterclockwise. As a result, these planes will spin in circles forever.

We decided to abandon the A3C approach because of three behaviors. First, 100 percent of all A3C training simulations ended in learning collapse. Every single training attempt eventually resulted in spinning circles by 3,000 iterations. The convergence rate on this behavior indicated that the way that AFGYM posed our planning problem was ill suited for A3C. Second, usable results had to be cherry-picked from a precollapse checkpoint. We analyzed all of the training checkpoints and discovered that the majority of them were ineffective at completing the mission. Finally, most of the A3C successes were because of reward-shaping. Because the majority of simulations did not work, we helped out the algorithm by assigning rewards for closing the distance to or pointing toward the target. The reward function was highly specialized and was approaching a rule-based system.

Figure 3.2. A3C Algorithm Results

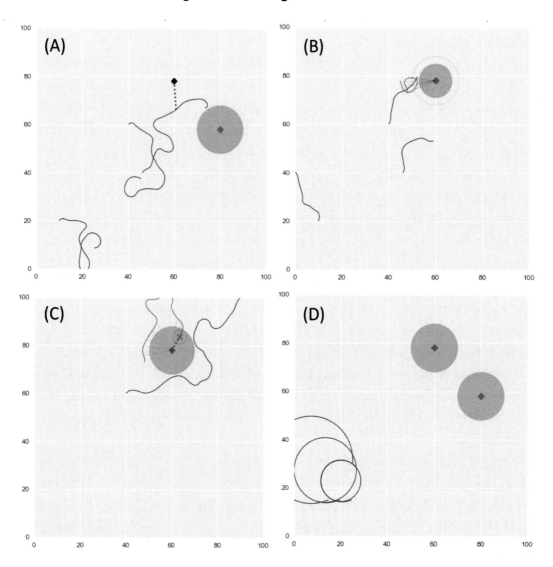

We developed a single-agent algorithm via PPO that was robust to changes in the laydown. Trained algorithms demonstrated robust and generalizable behavior across different laydowns. Of 10,000 simulations in AFGYM, this algorithm succeeded 96 percent of the time. Figure 3.3 shows the laydown. Panel (A) is a depiction of the variables that are varied in each laydown. The fighter's initial position is randomized within the light blue square. Its missile range and initial heading are also varied. The SAM and the target are also varied within their respective squares. The SAM is initialized with variable ranges as well. Panel (B) shows an example of a successful mission controlled by the PPO-trained algorithm. In this snapshot, the fighter has just eliminated the target while avoiding the SAM safely. Movement is much less erratic than in any of the panels in Figure 3.2 because the fighter takes a more direct approach toward its target. Because of the randomizing of ranges, the fighter outranges the SAM in 35 percent of the scenarios, so the algorithm preferred to avoid the SAM in most cases (only 3 percent of the 10,000 simulations resulted in the elimination of both the SAM and the target).

Figure 3.3. PPO Algorithm Results

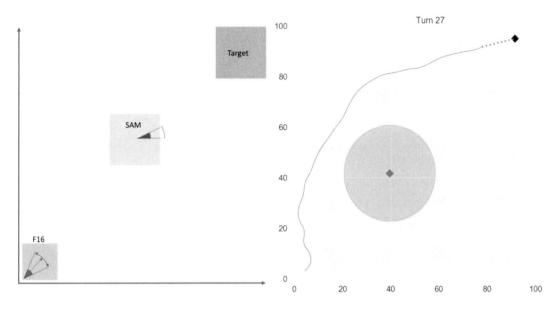

We also trained a multiagent scenario involving a fighter and a decoy. Figure 3.4 illustrates this scenario. Panel (A) shows a target and a SAM that is protecting it. The fighter and decoy are starting at different areas. Panel (B) shows the decoy being eliminated by the SAM. The SAM has a short delay before it can fire again. Panel (C) shows that the fighter takes advantage of this delay by eliminating both the SAM and target. Modifications to the laydown (e.g., fighter and decoy starting positions and initial headings) illustrated that the algorithm had not fully learned how to time the Blue agents to fully leverage the SAM's delay. For example, in many failed missions, the fighter ingresses too quickly before the decoy. Interestingly, there were some missions where the decoy deliberately avoids the SAM. Of 10,000 AFGYM simulations where we modified the starting positions, there was an 18-percent success rate. In comparison, when starting positions were fixed and only initial headings were varied, the success rate was 80 percent. This algorithm was less robust than the single agent.

Figure 3.4. PPO Algorithm Results with Decoys

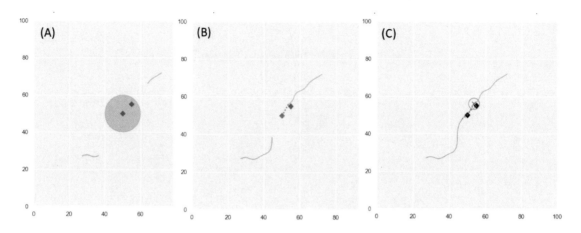

Finally, we demonstrated that the single-agent PPO algorithm succeeds in the AFSIM environment as well. Figure 3.5 shows that the algorithm successfully navigated an AFSIM laydown similar to that of Figure 3.3. The fighter navigates toward the SAM in an orthogonal fashion, destroys the SAM, and then ingresses toward the target. This exercise served as a proof of concept for our communication layer between the Python PPO algorithm and the AFSIM environment. This domain adaptation across two different environments shows that our trained agent transfers across to a different simulation than it was trained on. The robustness and extent of this transferability was not explicitly tested in this study because of the lengthy time each AFSIM simulation takes. However, we believe that the PPO algorithm should achieve similar success rates over 10,000 because the communication layer ensures that the PPO algorithm "experiences" the same environment input as that of AFGYM.

Figure 3.5. Demonstration of PPO AFGYM-AFSIM Transferability

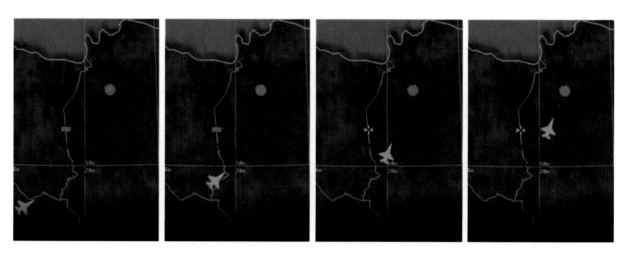

In contrast to A3C, PPO was superior in terms of performance. Learning collapses were rare and occurred about 5 to 10 percent of the time. Learning remained stable after thousands of iterations. Furthermore, PPO learned generalizable behaviors that were not possible with A3C. PPO agents demonstrated generalizability across different laydowns (96 percent) and even worked in AFSIM. As a comparison, A3C did not demonstrate any generalizability and its success rate in similar tests was less than 5 percent because the agents would fly into the SAMs or exit the battlefield. Furthermore, PPO learned robust SAM avoidance behavior by flying both above and below the SAM on its way to striking the target.

Possible Future Approaches to the 2-D Problem

Incorporating varying numbers of agents can be a challenge. The agents in our case are the two types of aircraft and the SAMs. In our current approach, we set up the input and output of the algorithm in a way that constrains the number of elements to five aircraft and five SAMs. Decreasing the number of agents is possible by simply placing the extra agents in places that are not relevant or by giving them zero values. However, increasing the number of agents is not possible without increasing the size of the input or output space and retraining the model. This way of arranging the agents and the corresponding neural network architecture is illustrated in Figure 3.6.

Figure 3.6. Depiction of Current Neural Network Representation of State and Outputs

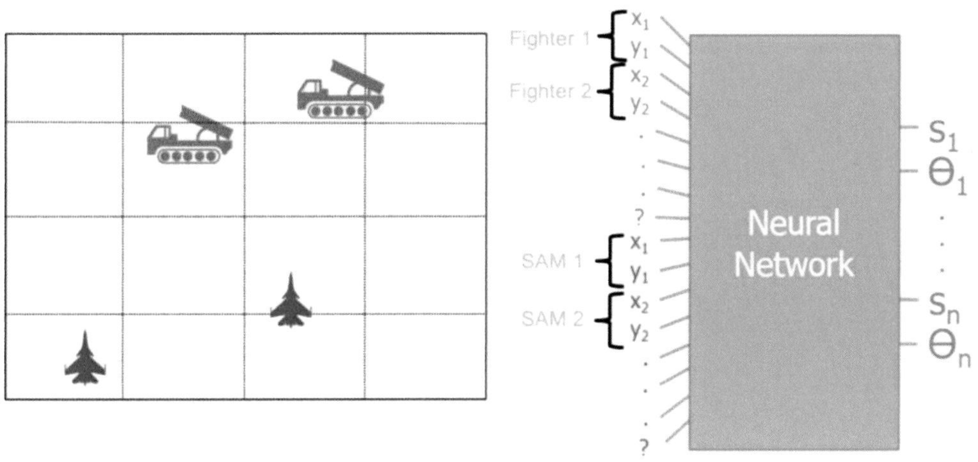

An alternative option for future exploration is to define the inputs by region and have the output correspond to a single decision. This alternative way of structuring the state and corresponding neural network is depicted in Figure 3.7. In the figure, x and y define the location of each agent where the subscript indicates which specific agent is being referred to and s and Θ indicate the speed and direction of the agents respectively.

Figure 3.7. Depiction of Geographically Based Representation of State and Outputs

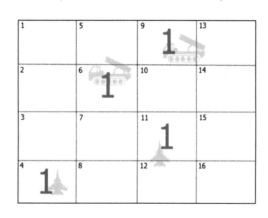
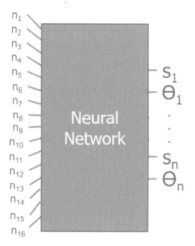

Structuring the state in this way makes it natural to incorporate varying numbers of agents, because each geographical region in the map can contain a varying number of agents of each type. Figure 3.8 illustrates the same map and architecture, but with 17 SAMs and 16 fighters.

Figure 3.8. Geographically Based Representations Allow for Arbitrary Numbers of Agents

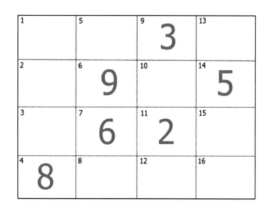
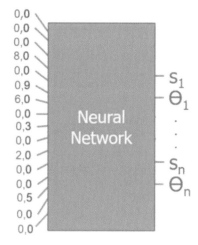

This way of describing the state has the advantage of being able to incorporate large numbers of agents naturally and is a possible path forward for future work in which large swarms are considered. The method does incur design trade-offs and limitations in exchange for potentially greater generality. The input vector to the network is now substantially larger and scales with the grid resolution of the space. The precision of the system laydown representation is also limited by the grid resolution.

4. Computational Infrastructure

A robust ML exposition of the mission-planning problem space could require substantial processing power and training time, constraining the utility of those systems. This section discusses the role of a computational infrastructure for the RL using AFSIM. Current deep RL methods are computationally intensive and require many evaluations to train.[55] For example, the nondistributed version of AlphaGo used 48 CPUs and eight graphical processing units (GPUs). The distributed version leveraged 1,202 CPUs and 176 GPUs. Tens to hundreds of millions of training steps were required for the policy and value networks. Likewise, generating sufficient volumes of training data takes substantial time and can be computationally intensive. Furthermore, the AlphaGo simulation was a gameboard, which requires minimal computational resources to run. The more complex simulation in the current problem adds to the computational burden.

An opportunity to build a flexible set of infrastructural components into the AI system was recognized through early discussions about the overall approach. With a thoughtful system design, there might be a natural, cost-effective way to scale the AI system for more-complex, time-intensive, or data-intensive implementations. Table 4.1 details the factors that are considered in computational infrastructure design.

Table 4.1. Examples of Factors Influencing Computational Infrastructure

Factor	Potential Considerations
Parallelism of the algorithm	To what extent is the algorithm parallelizable? If the algorithm is intrinsically parallel, how many computational workers are minimally needed and how are their activities coordinated?
Reliance on large data sets	Per iteration, how many data will be used by the algorithm? Is this data local or externally derived? How many data need to be exchanged with other components or entities, and at what frequency?
Modularity and portability of the algorithm with respect to implementation	Can individual components of the system be updated and deployed without requiring updates to the entire system? What development pipeline is required to support the different research paths? How easy is it to deploy new versions?
Algorithmic complexity	How does computational complexity scale with problem size? How well does the computational platform inherently scale to support the problem being researched?
Methods for component-to-component interaction	If components within the system interact, which communication mechanisms need to be supported? To what extent are network or web-based services used? Are standardized interface definitions available?

---

[55] David Silver, Aja Huang, Chris J. Maddison, Arthur Guez, Laurent Sifre, George van den Driessche, Julian Schrittwieser, Ioannis Antonoglou, Veda Panneershelvam, Marc Lanctot, Sander Dieleman, Dominik Grewe, John Nham, Nal Kalchbrenner, Ilya Sutskever, Timothy Lillicrap, Madeleine Leach, Koray Kavukcuoglu, Thore Graepel, and Demis Hassabis, "Mastering the Game of Go with Deep Neural Networks and Tree Search," *Nature,* Vol. 529, January 27, 2016.

Factor	Potential Considerations
Degree of coupling between AFSIM and the learning module	To what extent do execution of AFSIM and the RL engine rely on technical details of the other component's implementation? Can each component be evolved and complete its work independently of the other? Are shared memory or shared data stores being leveraged?
Information safeguarding and security	Which tools and secondary functionality are required to safeguard information? How might existing security mechanisms constrain development for the research effort?

For scaling the computational needs of the final AI system, an initial focus was placed on preparing an environment for the AFSIM computing component, potentially as a cluster of computing systems.[56] We focused on private and public cloud offerings with a low point of entry (having immediate availability within our research setting). Although relevant for this work, an ability to distribute AFSIM onto a cloud architecture could also be quite useful for other analytic work.

Our next task was to detail the operational context within which AFSIM would execute. AFSIM's resource requirements scale with the size of the problem space and number of runs. For runs that were more computationally intensive, a method to launch multiple instances of AFSIM, either on a single computer node or multiple nodes would require additional development, as AFSIM does not currently support parallel or distributed execution natively.[57] Execution of AFSIM itself is straightforward and from the command-line. Scripting is still required to launch AFSIM; collect or analyze results; and relaunch AFSIM with new parameters, either on a scheduled basis or triggered externally. The application-level interface to the RL component remained open during the initial phase, as the model for the ML was made more precise.

Platform-agnostic solutions and management of the AFSIM instances were also considered, and a "container"-based approach seemed to satisfy both needs effectively within the scope of our project. Container-based management would (1) neutralize concerns about maintaining a consistent operating environment on any platform, (2) "spin up/down" AFSIM computing instances, and (3) potentially enable cluster-based techniques, such as load-balancing. AFSIM itself runs in Windows and Linux variants. Deployment using standard cloud-management and orchestration tools (e.g., Chef, Puppet, Ansible, CloudFormation) for machine "templates" would be more streamlined as well.[58] Appendix B details the process and challenges associated with the containerized deployment of AFSIM and the planning agent.

---

[56] We focus on AFSIM here, as the AFGYM utility was specifically designed to execute fast enough to not require a distributed computing environment.

[57] AFSIM supports integration to distributed interactive simulations and high-level architectures for distributed simulation contexts, but not for purely "computational" reasons. Additionally, although core development of the AFSIM implementation is possible to natively leverage cloud-based services versus cloud-*hosted* services, the development effort would be labor-intensive and out-of-scope for this project.

[58] A brief discussion of cloud management and orchestration tools is presented at Gruntwork blog (Gruntwork, "Why We Use Terraform and Not Chef, Puppet, Ansible, SaltStack, or CloudFormation," blog post, September 26, 2019.

In summary, a virtualized, proof-of-concept environment for ML and AFSIM execution based on highly portable containers was developed. The AFSIM 2.2 simulator, implementations of the ML engines using TensorFlow, and parts of a batch execution framework have been prepared in Docker containers on both the internal platform and the external Amazon Web Services GovCloud platform, with initial tests at least partially successful, particularly on the external architecture. The use of containerization supports more-efficient research experimentation and choice of algorithms through reuse of AFSIM containers and for a variety of ML techniques. This approach should be explored by analytic efforts in need of large-scale computing resources.

Computing Infrastructure Challenges

Standard, precautionary environmental controls and configuration of the default, private cloud platform represented the most-significant hurdles for development of the container-based solutions. Security-oriented controls impeded installation of containerization software and formation of base images alike. The Docker tool relies on public registries to make standard base images available, for example, and a high degree of communication is required with external services to form new images locally and to update Docker itself.

Although the allocation of disk space on the private cloud was sufficient for running an AFSIM experiment, it effectively restricted the "working" space required to efficiently manage and develop improved solutions, relying on multiple AFSIM instances. This constrained our ability to implement and test container-based images and also produce training data in a distributed fashion to accelerate testing. Although multiple virtual machines (VMs) were available, we found disk space to be shared within the scope of an account owner, raising potential conflicts as code was updated or data were produced. Significant platform-specific scripting would be required to mitigate the issue. Furthermore, the containerization tool was installed on a partition at the system level in a way that prevented our reconfiguring the tool to use available disk space in a flexible manner; attempts to reconfigure the tool were not successful. Shifting development to the public Amazon Web Services GovCloud platform resolved all of these issues and the standard tooling enhanced management of the AI system.

Interoperability, communication of data between components of the ML architecture, and inclusion of third-party dependencies also presented technical hurdles. Although interoperability and interface definitions are a standard challenge, the uncertainty in the long-term design of the ML architecture left open the best approach for a "stable" ML container definition (base image for the AI system). For example, best practice is to form multiple containers—one for each component of the application—with small footprints that are independently maintained. The interaction among these components can be coordinated through host system resources, such as shared disk volumes, or through service-oriented design patterns. As the system evolves, the base image or images used in real-world deployments might require very different support from the underlying computing infrastructure. However, the expectation is that these issues can be resolved within the currently selected cloud platform and using the defined approach for container preparation.

Considering this, the focus shifted to using the external GovCloud cloud platform. An Amazon Elastic Compute Cloud instance (EC2) running Ubuntu 16.04 was defined. A new Docker image was prepared with AFSIM 2.2, and a 2-D implementation of the RL system bundled together. In this case, the base image was expanded to include Python 3, TensorFlow, NumPy, and Pandas. As described in previous chapters, this AI system will exchange data between runs through file-passing (within the container), and results are accessible to the host system via a mounted volume using a standard process. Although this system is under development, the basic facility for a combined AFSIM and AI system has been established. Next steps include a full execution of the AI system for analysis, and use of larger EC2 instance templates for distributed ML experiments over multiple instances of the AI system. Reformation of the current multiapplication image into several smaller ones will enable the desirable level of flexibility and degree of freedom required to mature the system into a real-world application.

5. Conclusions

In this report, we explored ML approaches to the SEAD problem. We created AFGYM to simulate SEAD missions. AFGYM is a low-fidelity version of AFSIM and was designed to have fast runtimes for the purposes of RL research. We trained RL agents to perform increasingly complex SEAD missions. First, we started with fighter and fighter-jammer scenarios in 1-D. We tested Q-learning and GAN algorithms and discovered that they both performed well. Next, we transitioned into fighter, fighter-decoy, and fighter-jammer scenarios in 2-D. We tested A3C and PPO algorithms and discovered that only the latter worked well.

This project prototyped a proof-of-concept AI system to help develop and evaluate new CONOPs in a test problem of simplified strike mission planning. Given a group of UAVs with different sensor, weapon, decoy, and EW payloads, we attempted to build agents that could employ the vehicles against an isolated air-defense system. We explored several contemporary statistical learning techniques to train computational air-combat planning agents in two operationally relevant simulation environments. This included the development of a simplified environment to allow rapid algorithm development and integration of ML agents with the AFSIM combat simulation. Although the proposed test problem is in the air domain, we expect analogous techniques with modifications to be applicable to other operational problems and domains.

The project was originally conceived as a mission-package planner, but pivoted to an air-combat planner when we realized that we could not achieve the former without the latter. During the project, we realized that a fundamental limitation to neural networks is that they are universal function approximators; they map a fixed-size input to a fixed-size output. As a result, we trained our algorithms to control a fixed number of Blue agents. We did not have a chance to research a multiagent AI, which will be necessary for a mission-package composer. This setback was a limitation of our problem formulation and not RL. We removed multiagent training from the scope of the project to focus on algorithm testing and development.

This exploratory research highlighted both the potential of RL to tackle complex planning problems and some of the continued limitations and challenges to this approach. Specifically, pure RL algorithms can be inefficient and prone to learning collapse. The catch-22 with RL was that we needed large numbers of training iterations to develop a good agent, but large numbers of iterations made it more likely for learning to collapse. All algorithms that were used in the 2-D case suffered from learning collapse. PPO is a recent step in the right direction because of the built-in constraints preventing the network parameters from changing too much in each iteration. These constraints alleviate the manual hand-tuning of a heuristic learning rate parameter. Future work in this area can try to address the learning collapse problem in more-systematic ways. Some specific next steps in the research program could include:

- **Adding increased problem complexity and scale:** Adding more platforms would force the ML to learn coordination strategies between UAVs. Scaling up Red forces by adding multiple SAMs, targets, and terrain obstacles would make the environment more realistic. Moving into a 3-D space with terrain effects is also logical.

- **Focusing on the multiagent problem:** Future work on automated mission planning should focus on developing robust multiagent algorithms. Much inspiration can be drawn from recent StarCraft 2 or DOTA 2 AI bots, which are AI networks that control multiple characters concurrently. StarCraft AI has seen advances in bidirectional recurrent neural networks as a way for multiagents to communicate with each other. DOTA 2 AI has seen advances in developing independent bots that interact without communication and rely on "trust" to sacrifice short-term rewards for greater long-term rewards (winning the game). OpenAI's implementation of DOTA 2 bots appears to use an operator taken from image analysis (*maxpool*) to pool information from an indefinite number of agents.[59] Implementation of both techniques is ongoing but has proven to be difficult because no open-source implementations exist yet.

- **Demonstrating transfer learning from AFGYM to AFSIM AI:** In the 2-D problem, the agent was trained in the AFGYM environment. After training, by matching AFSIM and AFGYM's output formats at each time step, we successfully used the pretrained agent in coordination with the AFSIM environment (thereby removing AFGYM from the process). In the future, we would like to train the agent directly in the AFSIM environment, which would require using multiple instances of AFSIM in parallel, which, in turn, is a large computational burden. For analyses involving the use of large simulations in place of large datasets, the required computational burden will continue to be a significant challenge. This also raises the more general point that modeling and simulation systems and other "target" environments that can potentially provide feedback for learning will need to evolve with advances in ML.

- **Exploring improvements through mutual self-play:** Contemporary AIs, like Alpha Zero and the DOTA 2 bot, can learn and improve through mutual self-play at scale. However, because the current problem is asymmetric with dissimilar Blue and Red agents, a SAM ML agent will be needed to allow for adaptive self-play.

- **Automating hyperparameter turning:** Hyperparameter tuning is both important and undertheorized. Automated exploration of hyperparameters is key for preventing premature learning collapse.

- **Increasing the state representation and problem formulation:** The current representation of the environment is inflexible. The ML agent can be trained against a fixed number of agents and targets. If we introduce additional targets or agents, then the input vector size will grow, and we need to retrain the agent. In Chapter 4, we outlined one possible path forward to build a more generalized representation of the problem.

Policy Implications

Finally, we return to the DoD mission planning problem discussed in Chapter 1. The main challenge highlighted there was the complexity created by large numbers of combinations of

---

[59] "OpenAI Five Model Architecture," webpage, June 6, 2018.

platforms, sensors, and other capabilities that could exceed human abilities to plan and coordinate effectively in a timely fashion. Obviously, our demonstrations in Chapter 2 and Chapter 3 did not fully capture this level of complexity. However, our success in using AI systems to develop time- and space-coordinated flight routes under different initial conditions indicate that this approach, when approached at scale and with better tuning, should have utility. And, given that state-of-the-art planners do not interactively coordinate the actions of multiple platforms, an integrated planning capability would mark a significant improvement. However, the possibilities come with the following important caveats:

- The scaling of computational power and time required to train realistic sets of capabilities (dozens of platforms) against realistic threats (dozens of SAMs) remains unclear. In peacetime, months and hundreds of computers might be available, but dozens of years or tens of thousands of computers are not. However, once training is complete, the actual planning process should be quite fast, compared with today's automated planners that can take hours to evaluate a complicated mission, followed by further hours of manual fine-tuning. Adapting current ML approaches to mission planning creates demand signals for developing fast, scalable, and flexible modeling and simulation environments.

- Introduction of new categorical capabilities by Blue or Red that are not simply changes in range, for example, would require retraining the algorithm to take advantage of—or avoid—them. If realistic algorithm training takes weeks on a supercomputer cluster, this might not be usable during a conflict. Nevertheless, our results indicate that trained algorithms should be able to deal with changes in number and locations of assets fairly easily. The longer-term solution to the problem of generalization lies in fundamental advances in modular and composable networks, transfer learning, and hybrid methods that integrate probabilistic learning with knowledge representation so that we can leverage prior knowledge and avoid having to start each time from tabula rasa.

- The algorithms explored here can easily produce results that are unexpected and might be unacceptable to humans. Thus, for example, situations could arise where a few aircraft are sacrificed to save a larger number. Even in combat situations, this outcome is usually not acceptable. In actual conflicts, this effect is captured as an acceptable level of risk where some missions are more important than others, and so more risk is acceptable; however, there is no risk "knob" in the framework that we developed. To force an algorithm to respect this, increasing risk would need to be captured as a penalty on the mission score during training.

- Few real-world data exist on successful and unsuccessful missions. Compared with the volumes of data used to train contemporary AI systems, very few real missions have been flown against air defenses, and virtually all of them were successful. Recall that we required training sets in the tens of thousands for even our simple cases. Thus, simulations will be needed to train these types of algorithms, and simulations do not perfectly represent the real world. There is a risk that algorithms, trained to perfection against the simulated world, will behave very poorly in the real one. As seen with the development of autonomous vehicles, developing trust will require more-exhaustive testing and fundamental advances in algorithm verifiability, and safety and boundary assurances.

These technical and transition challenges have profound policy implications. Today, many parts of the U.S. government are investing large amounts of money and numbers of personnel in exploring applications for AI, particularly ML. For some areas, where significant labeled data are available and the problems are like those the algorithms were developed to solve (recognition of features in intelligence imagery, for example), fairly rapid success is likely.

However, for applications lacking these two attributes, progress could be slow. For example, as pointed out previously, using simulations to produce training data brings with it the whole set of computational and algorithmic challenges. The particular problem explored here demonstrated some progress, but only for cases simple enough to be manually planned. Clearly, significantly more investment in data generation and training time would be required to handle more-complicated situations. For applications involving strategic decisionmaking, such as those where simulations do not even have physics to fall back on, there might be so little correspondence between the real world and the simulation that trained algorithms will be effectively useless. Furthermore, the end objective of high-level strategic reasoning is often to undermine or change the rules of the game rather than to optimize for best outcome within the constraints. National security decisionmakers will have to be judicious about focusing investments in areas likely to bear fruit, leaving others as topics for more basic research. Not every security problem will be solved by AI.

Appendix A. 2-D Problem State Vector Normalization

All elements of the 2-D state vector, which is updated for the agent after every time step in AFSIM, contains values normalized between 0 and 1. This normalization of the input space helps with convergence and stability. The normalization is outlined in Table A.1. The scenario laydown is set up for normalization by considering the target's and fighter's initial locations as marking opposite points on a rectangle. All events in the scenario take place in this rectangle, and the lengths of this rectangle are used to normalize several scenario parameters as described in the next section. In Table A.1, the *FighterStart* variable refers to the fighter's randomly generated starting point, which serves as an anchor for many of the normalization calculations below. We also use *TotalLatLength* and *TotalLongLength* as normalization factors which are defined as:[60]

$$TotalLatLength = Target_{Lat} - FighterStart_{Lat} + 0.1$$
$$TotalLongLength = Target_{Long} - FighterStart_{Long} + 0.1$$

---

[60] The additional 0.1 ensures that, when normalized by *TotalLenLength* or *TotalLongLength*, we get values slightly less than 1, so the fighter has the opportunity to reach the target without hitting the scenario's boundary, at which point the ML automatically stops.

Table A.1. 2-D Problem Formulation—State Reported to ML Agent at Each Time Step

State Variable	Normalization
Fighter heading	Degree orientation used in AFSIM is first mapped onto standard unit-circle degree measure (i.e., 0 degrees = due east, 90 degrees = due north). We call this transformed degree heading *FighterHeading*. To normalize, we take *FighterHeading/360 (Fighter*$*(\pi/180)/2\pi$) which is in [0,1]
Fighter latitude	*(FighterCurrent_lat – FighterStart_lat)/TotalLatLength*
Fighter longitude	*(FighterCurrent_long – FighterStart_long)/TotalLongLength*
Fighter dead	1 if dead, 0 if alive
Fighter firing range	The firing range, prenormalization, is in meters. We normalize with respect to the sum of the laydown rectangle's length and height, which is also in meters: *FighterFiringRange(normalized) = FighterFiringRange/ [(TotalLatLength)*40008000/360 + TotalLongLength*(40075160*Cos(FighterStart_lat)/360)]*[a]
Fighter weapon count	*FighterWeapCount(normalized) = FighterWeapCount/max(SAMWeapCount, FighterWeapCount)*
SAM heading	0 – fixed report to agent
SAM latitude	*(SAMCurrent_lat – FighterStart_lat)/TotalLatLength*
SAM longitude	*(SAMCurrent_long – FighterStart_lat)/TotalLongLength*
SAM dead	1 if dead, 0 if alive
SAM firing range	*SAMFiringRange(normalized) = SAMFiringRange/[(TotalLatLength)*40008000/360 + TotalLongLength*(40075160*Cos(FighterStart_lat) /360)]*[b]
SAM weapon count	*SAMWeapCount(normalized) = SAMWeapCount/max(SAM WeapCount, FighterWeapCount)*
Target heading	0 – fixed report to agent
Target latitude	*(TargetCurrent_lat – FighterStart_lat)/TotalLatLength*
Target longitude	*(TargetCurrent_long– FighterStart_long)/TotalLongLength*
Target dead	1 if dead, 0 if alive

[a] The coefficients on *TotalLatLength* and *TotalLongLength* are standard conversions for mapping latitude and longitude units into meters. In the second term, we take the cosine of the *FighterStart_lat* for simplicity.
[b] The coefficients on *TotalLatLength* and *TotalLongLength* are standard conversions for mapping latitude and longitude units into meters. In the second term, we take the cosine of the *FighterStart_lat* for simplicity.

Appendix B. Containerization and ML Infrastructure

Containers extend virtualization beyond machines to individual applications, enabling them to be run on any platform. A container is the running instance of an application that has been "imaged" in a way that permits it to be executable on any computer. VMs virtualize the physical computing environment used by applications and a broad variety of operating systems. Containers go a step further by allowing an application prepared to run in one host's environment to be executed as-is by a different operating system and host computer. Figure B.1 illustrates the differences between containers and VMs with respect to applications. In particular, containers do not interact with the operating system directly. Effectively, containers virtualize operating systems by (a) packaging all of the required application components into an image, and (b) bridging execution of the application's image with the current host's operating system.[61]

Containers are similar to VMs and also offer several advantages. Both constructs for application delivery and experimental environments are cloud-compatible, provide a reproducible setup, and offer bundled, self-sufficient deployable packages for the research product. The application-centric point of view applied by containers introduces several additional advantages that are also compatible with use by VMs and improves the overall utility of physical infrastructure. Because containers represent running applications instead of an entire machine of applications, the requirements to prepare, test, and deploy an application are dramatically reduced. Containerization maintains and bridges the application to its external dependencies (including other containers) for applications that are required by the system to operate. A general comparison of VM and containerized applications is presented in Figure B.1 and Table B.1.

Figure B.1. Containers and Virtual Machines

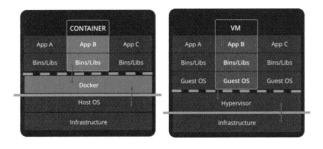

SOURCE: Based on Docker, "Orientation and Setup," webpage, undated.

---

[61] Historically, containerization dates back to the late 1970s, with a simple mechanism to change the base reference point used by an application to access to resources (its "root" for the file system). The relationship to virtualization and its motivation is described in numerous texts. In the early 2000s, containerization matured rapidly, and, by 2013, its regular use grew rapidly.

At the application level, such "platform independence" facilitates an architecture for reproducible computational research and portability of the research product. The expectation is that the overall return on investment is improved by lowering development and operational costs and enabling use of a broad variety of potentially heterogeneous platforms and tools for real-world applications and use in decisionmaking. Standard and publicly available tools to coordinate or "orchestrate" execution of a group of containerized applications provide another mechanism to boost the utility of the selected infrastructure and dynamically address scalability as, or when, the AI system might require. Architectural design patterns utilizing open, modular, or service-oriented methods (such as microservices) are also enabled.

Table B.1. A Brief Comparison of Virtual Machines and Containerized Applications

	Virtual Machine Virtualize the Computer	Container Virtualization for Applications
Bundled, consistent, self-sufficient runtime environments	X	X
Cloud-ready	X	X
Platform for reproducible research	X	X
Minimal installation and execution overhead		X
Fast spin-up		X
Simple and fast updates and maintenance		X

Viewing the problem from the point of view of the application and the resources it specifically needs to operate, we took advantage of containerization to package AFSIM batch execution in a self-sufficient manner and to expand options for deployment and scalability without extensive effort.

Because of its popularity, degree of cross-platform support, and low entry point, Docker was chosen as the containerization tool for our research.[62] Docker also provides a private, local registry service that facilitates distribution of containers within developer teams. It is worth noting that high-performance computing (HPC) environments might provide or require specialized containerization tools to support research on HPC systems.[63] In fact, containerization is being directly applied to deep learning research in HPC environments at Oak Ridge National Labs to support research on cancer behavior and treatment—specifically for its ability to make the latest ML tools available, in a standard way, to a research community—and for its capacity to support algorithmic diversity in this domain.[64]

---

[62] Karl Matthias and Sean Kane, *Docker Up and Running*, Sebastopol, Calif.: O'Reilly Media, 2015.

[63] For example, see Sylabs.io, "Documentation and Examples," webpage, undated; "'Shifter' Makes Container-Based HPC a Breeze," blog post, National Energy Research Scientific Computing Center, August 11, 2015; and Joab Jackson, "Containers for High Performance Computing," blog post, *New Stack*, June 26, 2018.

[64] Rachel Harken, "Containers Provide Access to Deep Learning Frameworks," blog post, Oak Ridge National Laboratory, undated; Kathy Kincade, "A Scalable Tool for Deploying Linux Containers in High-Performance Computing," blog post, Lawrence Berkeley National Laboratory, September 1, 2015.

Initially, Docker was used to create an image of AFSIM 2.0.1 on a laptop, with AFSIM scripts being executed to generate a large volume of simulation results. The containerization was itself performed inside of a virtual machine prepared with an open-source Linux variant as its operating system (Ubuntu 16.04), and included a reduced-footprint Python 2.7 image as a starting point. Once this example system was determined to function as intended, multiple servers running Red Hat Enterprise Linux on the RAND internal cloud were prepared with Docker. This private cloud version of Docker was then used to prepare AFSIM 2.2 images and generate simulation data for training the current ML implementation on the 1-D scenario. The container-generated AFSIM simulation results were subsequently passed to a separate RL application that was actively under development on a separate system and being managed manually.

However, the design of the private cloud operating environment limited our ability to prepare images efficiently and to coordinate the work performed by multiple containers on different virtual hosts. The Actor-Critic model being explored would require multiple ML workers with AFSIM in the loop, and potentially many more training iterations. Furthermore, the parallel generation of larger volumes of training data further required custom coding to manage a system-specific disk configuration in the private cloud, which was counter to the goals for the flexible computing infrastructure.

Appendix C. Managing Agent-Simulation Interaction in the 2-D Problem

The information flow between the AFSIM simulation environment and the Python-based RL agent is illustrated in Figure C.1. The approach is common for different RL agents.

Figure C.1. AFSIM/ML Agent Coordination

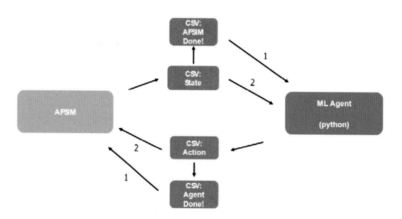

NOTE: CSV refers to a comma-separated value text file.

The following list summarizes the key steps:

1. The run begins with the ML agent "listening" for AFSIM's first state report.[65]
2. The AFSIM scenario begins with the fighter flying directly toward the target.
3. After the fighter has flown a total distance of 0.5 units in normalized space (see Appendix A for normalization details), AFSIM reports its first state update into a text file, which we refer to as the "State" file.
4. This triggers the creation of another text file to act as a flag for the agent. This file—the "AFSIM-Done.csv" file—signals that a new current state has been appended to the State file.[66]
5. AFSIM goes into "listening" mode, awaiting a flag that its next action is ready from the agent.
6. The agent removes the newly created "AFSIM-Done.csv" file from its directory.

---

[65] *Listening* here refers to the agent's checking every second whether an "AFSIM done" file has been created in its current file directory, which signals that AFSIM's newest state is ready. This is described in later steps of the process.

[66] Appending data to the State file cannot itself act as the flag that a new state is ready. When we tried this approach, we found the behavior to be erratic, often because the agent (or AFSIM, when acting in reverse) would detect that the State (or Action) file had been appended when only a fragment of the data had been populated.

7. Based on the latest state vector fed to the agent, the agent returns an action to an "Action" file. In our problem, the action only consists of a new heading change for the fighter.
8. The agent creates an "Agent-Done.csv" file in the working directory. This acts the signal for AFSIM that a new action (fighter heading) is ready.
9. AFSIM removes the newly created "Agent-Done" file from the directory.
10. The fighter is directed to turn toward the latest heading and proceed another 0.5 units (in normalized space).

The agent and AFSIM pair continue in this loop until the simulation completes or the fighter is shot down.

Appendix D. Overview of Learning Algorithms

MDP Formulation

The tuple $(S, A, P_a(s, s'), R_a(s, s'))$ describes the MDP formulation of Figure 1.1. S is the set of all possible states an agent can experience within the environment. A is the set of all possible actions an agent can take within the environment. $P_a(s, s')$ is a Markov transition matrix describing the probability of the agent moving from any state s to another state s' via action a. $R_a(s, s')$ is a reward matrix describing the expected reward for transitioning from any state s to another state s' via action a. The goal of an RL algorithm is to identify an optimal policy, π^*, that maximizes the expected rewards over time. A discount factor, $\gamma < 1$, is used to stabilize the series over an infinite timespan.

$$\pi^*(s, a) = \underset{\pi}{\text{argmax}} \, \mathbb{E} \left[\sum_{t \geq 0}^{\infty} \gamma^t r_t | s_0 = s, a_0 = a, \pi \right]$$

Q-Learning

The following equations describe the foundations of Q-learning. The Q-function calculates the expected reward of taking action a at state s by taking the expected value of the reward (for performing the action) and adding the expected value of all future rewards (assuming you follow the current policy, π). The goal of Q-learning is to learn and approximate this Q-function for all possible states and actions. First, the problem is approximated by an MDP as follows:

$$(S, A, R, \boldsymbol{P}, \gamma)$$

S represents the set of all possible states in the environment, A represents the set of all possible actions, R represents a matrix encoding the reward an agent receives for all possible state-action pairs, P represents the transition probability from any state to another, and γ represents a discount factor. Specifically, we are using Q-learning to learn the ingress distance (action) for each platform, given 1-D laydown (state).

The Q-function is calculated as follows:

$$Q = \boldsymbol{E} \left[\sum_{t \geq 0} \gamma^t r_t | s_0 = s, a_0 = a, \pi \right]$$

This function can be interpreted as the total accumulated reward from interacting with an environment. The γ weighs future rewards less than current rewards and serves as a way to converge this infinite series. Interactions are defined as starting at state s_0 and taking initial action a_0 and then following policy π for the rest of the interaction. Because MDPs are probabilistic and actions are drawn from probability distribution $\pi(s)$, an expected value is wrapped around the entire term. Q-learning is an off-policy algorithm and seeks to approximate and map out Q over the entire environment, rather than learn π directly. The loss function for Q-learning, J, is to minimize:

$$J = E\left[(y - Q(s, a))^2\right]$$

$$y = E\left[r + \gamma \max_{a'} Q(s', a')|s, a\right]$$

Q represents the neural network's Q-function approximator and y represents an empirically calculated Q value (averaged over several batches of agent-environment simulations). The loss function is a sum of squared errors between the Q-function approximator and the real Q value calculated over many experiences. The y term is an iterative calculation that is used to select the most rewarding action in the next state, s'. The update step for Q-learning is:

$$Q(s, a) \leftarrow Q(s, a) + \alpha\left(r + \gamma \max_{a'} Q(s', a')\Big|s, a\right)$$

The α term is the algorithmic learning rate hyperparameter.

Generative Adversarial Networks

The CGAN can be interpreted as two networks competing against each other. One network (denoted by G) is designed to generate "fake" plans, given a scenario (*fake* defined as non–human generated). The other network (denoted by D) is designed to discriminate between real and fake plans conditioned on a scenario. The objective function, V, is designed as a minmax problem as follows:

$$\min_{G} \max_{D} V(D, G) = E_{x \sim p_{data}(x)}[log\ D(x|y)] + E_{z \sim p_z(z)}[log\ 1 - D(G(z|y))]$$

The x term represents plans, the y term represents the scenario that plans are conditioned on, and z represents a fake plan that was sampled from the probability distribution p_z. In practice, this minmax problem is very difficult to optimize. Training a GAN involves freezing one subnetwork to train the other. Specifically, the discriminator subnetwork is frozen, and the generator is trained to fool the discriminator ($\min_{G} V(D, G)$ is performed). Then, the generator subnetwork is frozen, and the discriminator is updated to discriminate between generated plans and real plans ($\max_{D}(D, G)$ is performed). This process repeats until a desired generator or discriminator performance is achieved.

A3C

A3C builds on the REINFORCE algorithm, which was designed as an on-policy algorithm. In contrast to Q-learning, which effectively amounts to the neural network storing a large table of state-action pairs, on-policy algorithms use a neural network to learn both the Q-function, $Q(s,a)$, and the policy function, $\pi(s)$, directly. Given the Q-function from before:

$$Q(s, a) = E[r + \gamma max_{a'} Q(s', a')|s, a]$$

$$J(\theta) = \boldsymbol{E}[Q(s,a)\log(\pi_\theta)\,(a_t|s_t)]$$
$$\nabla_\theta J(\theta) = \sum_{t \geq 0} Q(s_t, a_t)\, \nabla_\theta \log(\pi_\theta)\,(a_t|s_t)$$

The loss function for on-policy algorithms are defined as maximizing Q (the value of taking action a at state s) multiplied by the policy's probability of selecting action a given state s. The goal is to maximize this for every state-action pair in the environment. The update step is defined by the gradient of the loss function. Intuitively, if the Q value was high for a given state-action, then this update step increases the probability of taking that state-action in the future. If the Q value was low for a given state-action, then this update step decreases the probability of taking that state-action in the future.

First, A3C uses a computed value called *advantage*, which is defined as the difference between the machine's expected reward for taking an action and the actual reward received. The algorithm maintains a neural network that learns the value of taking different actions in a given state. The following equation shows the update rule for A3C. The above equations are slightly modified to change the Q term into $(Q - V)$ as follows:

$$J(\theta) = \boldsymbol{E}[[Q(a,s) - V(s,\theta)]\log(\pi_\theta)\,(a_t|s_t)]$$
$$\nabla_\theta J(\theta) = \sum_{t \geq 0} [Q(a_t, s_t) - V(s,\theta)]\, \nabla_\theta \log(\pi_\theta)\,(a_t|s_t)$$

PPO

TRPO limits the KL between the pre-update π_θ and the post-update $\pi_{\theta'}$. This constrained optimization is intractable to calculate in practice, and the TRPO method represents one such estimation using a surrogate objective function. We summarize the concept here, but refer the reader to the original paper for an in-depth explanation of TRPO.[67]

$$\pi_{\theta'} = argmax_\theta\, J(\theta)$$
$$s.t.\, KL(\pi_{\theta'}, \pi_\theta) \leq \epsilon$$

---

[67] Schulman et al., 2015; Schulman et al., 2017.

References

Arulkumaran, Kai, Marc Peter Deisenroth, Miles Brundage, and Anil Anthony Bharath, "A Brief Survey of Deep Reinforcement Learning," *IEEE Signal Processing Magazine*, Vol. 34, No. 6, November 2017, pp. 26–38.

Åström, K. J., "Optimal Control of Markov Processes with Incomplete State Information," *Journal of Mathematical Analysis and Applications*, Vol. 10, No. 1, February 1965, pp. 174–205.

Bahdanau, Dzmitry, Kyunghyun Cho, and Yoshua Bengio, "Neural Machine Translation by Jointly Learning to Align and Translate," *arXiv*:1409.0473, last revised May 19, 2016.

Bai, Haoyu, David Hsu, Mykel J. Kochenderfer, and Wee Sun Lee, "Unmanned Aircraft Collision Avoidance Using Continuous-State POMDPs," in Hugh Durrant-Whyte, Nicholas Roy, and Pieter Abbeel, eds., *Robotics: Science and Systems VII*, Cambridge, Mass.: MIT Press, 2012.

Brown, Jennings, "Why Everyone Is Hating on IBM Watson—Including the People Who Helped Make It," *Gizmodo*, August 10, 2017.

Canon, Scott, "Inside a B-2 Mission: How to Bomb Just About Anywhere from Missouri," *Kansas City Star*, April 6, 2017.

Chan, Dawn, "The AI That Has Nothing to Learn from Humans," *The Atlantic*, October 20, 2017.

Clive, Peter D., Jeffrey A. Johnson, Michael J. Moss, James M. Zeh, Brian M. Birkmire, and Douglas D. Hodson, "Advanced Framework for Simulation, Integration and Modeling (AFSIM)," *Proceedings of the International Conference on Scientific Computing*, 2015.

Defense Advanced Research Projects Agency, "DARPA Announces $2 Billion Campaign to Develop Next Wave of AI Technologies," webpage, September 7, 2018. As of December 17, 2019:
https://www.darpa.mil/news-events/2018-09-07

Defense Science Board, *Report of the Defense Science Board Summer Study on Autonomy*, Washington, D.C.: Office of the Under Secretary of Defense for Acquisition, Technology, and Logistics, June 2016.

Docker, "Orientation and Setup," webpage, undated. As of December 20, 2019:
https://docs.docker.com/get-started/#containers-and-virtual-machines

Ernest, Nicholas, David Carroll, Corey Schumacher, Matthew Clark, Kelly Cohen, and Gene Lee, "Genetic Fuzzy Based Artificial Intelligence for Unmanned Combat Aerial Vehicle Control in Simulated Air Combat Missions," *Journal of Defense Management*, Vol. 6, No. 1, March 2016.

Freedman, David H., "A Reality Check for IBM's AI Ambitions," blog post, *MIT Technology Review*, June 27, 2017. As of December 17, 2019:
https://www.technologyreview.com/s/607965/a-reality-check-for-ibms-ai-ambitions/

Frisk, Adam, "What is Project Maven? The Pentagon AI Project Google Employees Want Out Of," *Global News*, April 5, 2018. As of December 17, 2019:
https://globalnews.ca/news/4125382/google-pentagon-ai-project-maven/

Gillott, Mark A., *Breaking the Mission Planning Bottleneck: A New Paradigm*, Maxwell Air Force Base, Ala.: Air Command and Staff College, AU/ACSC/099/1998-04, April 1, 1998. As of December 17, 2019:
http://www.dtic.mil/dtic/tr/fulltext/u2/a398531.pdf

Gilmer, Marcus, "IBM's Watson Is Making Music, One Step Closer to Taking Over the World," *Mashable*, October 24, 2016.

Goerzen, C., Z. Kong, and Berenice F. Mettler May, "A Survey of Motion Planning Algorithms from the Perspective of Autonomous UAV Guidance," *Journal of Intelligent and Robotic Systems*, Vol. 57, No. 1-4, January 2010, pp. 65–100.

Goodfellow, Ian, Jean Pouget-Abadie, Mehdi Mirza, Bing Xu, David Warde-Farley, Sherjil Ozair, Aaron Courville, and Yoshua Bengio, "Generative Adversarial Nets," in Z. Ghahramai, M. Welling, C. Cortes, N. D. Lawrence, and K. Q. Weinberger, eds., *Advances in Neural Information Processing Systems (NIPS 2014)*, Vol. 27, 2014, pp. 2672–2680.

Gruntwork, "Why We Use Terraform and Not Chef, Puppet, Ansible, SaltStack, or CloudFormation," blog post, September 26, 2019. As of December 17, 2019:
https://blog.gruntwork.io/why-we-use-terraform-and-not-chef-puppet-ansible-saltstack-or-cloudformation-7989dad2865c

Harken, Rachel, "Containers Provide Access to Deep Learning Frameworks," blog post, Oak Ridge National Laboratory, undated. As of December 17, 2019:
https://www.olcf.ornl.gov/2017/05/09/containers-provide-access-to-deep-learning-frameworks/

He, Kaiming, Xiangyu Zhang, Shaoqing Ren, and Jian Sun, "Deep Residual Learning for Image Recognition," *Proceedings of the 29th IEEE Conference on Computer Vision and Pattern Recognition: CVPR 2016*, 2016, pp. 770–778.

Heckman, Jory, "Artificial Intelligence Vs. 'Snake Oil:' Defense Agencies Taking Cautious Approach Toward Tech," blog post, *Federal News Network*, December 12, 2018. As of December 17, 2019:
https://federalnewsnetwork.com/defense-main/2018/12/ai-breakthroughs-versus-snake-oil-defense-agencies-taking-cautious-approach-toward-tech/

Jackson, Joab, "Containers for High Performance Computing," blog post, *New Stack*, June 26, 2018. As of December 17, 2019:
https://thenewstack.io/roadmap-containers-for-high-performance-computing/

Jones, Jimmy, "System of Systems Integration Technology and Experimentation (SoSITE)," blog post, Defense Advanced Research Projects Agency, undated. As of December 17, 2019:
https://www.darpa.mil/program/system-of-systems-integration-technology-and-experimentation

Kania, Elsa, "AlphaGo and Beyond: The Chinese Military Looks to Future "Intelligentized" Warfare," blog post, *Lawfare*, June 5, 2017. As of December 17, 2019:
https://www.lawfareblog.com/alphago-and-beyond-chinese-military-looks-future-intelligentized-warfare

Kim, Yoochul, and Minhyung Lee, "Humans Are Still Better Than AI at Starcraft—For Now," blog post, *MIT Technology Review*, November 1, 2017. As of December 17, 2019:
https://www.technologyreview.com/s/609242/humans-are-still-better-than-ai-at-starcraftfor-now/

Kincade, Kathy, "A Scalable Tool for Deploying Linux Containers in High-Performance Computing," blog post, Lawrence Berkeley National Laboratory, September 1, 2015. As of December 17, 2019:
https://phys.org/news/2015-09-scalable-tool-deploying-linux-high-performance.html

Markoff, John, "Computer Wins on 'Jeopardy!': Trivial, It's Not," *New York Times*, February 16, 2011.

Martin, Jerome V., *Victory From Above: Air Power Theory and the Conduct of Operations Desert Shield and Desert Storm*, Miami, Fla.: University Press of the Pacific, 2002.

Matthias, Karl, and Sean Kane, *Docker Up and Running*, Sebastopol, Calif.: O'Reilly Media, 2015.

Mehta, Aaron, "DoD Stands Up Its Artificial Intelligence Hub," blog post, *C4ISRNET*, June 29, 2018. As of December 17, 2019:
https://www.c4isrnet.com/it-networks/2018/06/29/dod-stands-up-its-artificial-intelligence-hub/

Mirza, Mehdi, and Simon Osindero, "Conditional Generative Adversarial Nets," *arXiv*: 1411.1784, 2014.

Mnih, Volodymyr, Adrià Puigdomènech Badia, Mehdi Mirza, Alex Graves, Timothy P. Lillicrap, Tim Harley, David Silver, and Koray Kavukcuoglu, "Asynchronous Methods for Deep Reinforcement Learning," in Maria Florina Balcan and Kilian Q. Weinberger, eds., *Proceedings of the 33rd International Conference on Machine Learning*, Vol. 48, 2016, pp. 1928–1937.

Mnih, Volodymyr, Koray Kavukcuoglu, David Silver, Alex Graves, Ioannis Antonoglou, Daan Wierstra, and Martin Riedmiller, "Playing Atari with Deep Reinforcement Learning," *arXiv*: 1312.5602, 2013.

Mnih, Volodymyr, Koray Kavukcuoglu, David Silver, Andrei A. Rusu, Joel Veness, Marc G. Bellemare, Alex Graves, Martin Riedmiller, Andreas K. Fidjeland, Georg Ostrovski, Stig Petersen, Charles Beattie, Amir Sadik, Ioannis Antonoglou, Helen King, Dharshan Kumaran, Daan Wierstra, Shane Legg, and Demis Hassabis, "Human-Level Control Through Deep Reinforcement Learning," *Nature*, Vol. 518, February 25, 2015, pp. 529–533.

Nair, Arun, Praveen Srinivasan, Sam Blackwell, Cagdas Alcicek, Rory Fearon, Alessandro De Maria, Vedavyas Panneershelvam, Mustafa Suleyman, Charles Beattie, Stig Petersen, Shane Legg, Volodymyr Mnih, Koray Kavukcuoglu, and David Silver, "Massively Parallel Methods for Deep Reinforcement Learning," *arXiv*: 1507:04296, last updated July 16, 2015.

OpenAI, "OpenAI Five," blog post, *OpenAI Blog,* June 25, 2018. As of December 17, 2019: https://blog.openai.com/openai-five/

"OpenAI Five Model Architecture," webpage, June 6, 2018. As of December 20, 2019: https://d4mucfpksywv.cloudfront.net/research-covers/openai-five/network-architecture.pdf

Pfau, David, and Oriol Vinyals, "Connecting Generative Adversarial Networks and Actor-Critic Methods," *arXiv*: 1610.01945, last updated January 18, 2017.

Qiu, Ling, Wen-Jing Hsu, Shell-Ying Huang, and Han Wang, "Scheduling and Routing Algorithms for AGVs: A Survey," *International Journal of Production Research*, Vol. 40, No. 3, 2002, pp. 745–760.

Schulman, John, Sergey Levine, Philipp Moritz, Michael Jordan, and Pieter Abbeel, "Trust Region Policy Optimization," in Francis Bach and David Biel, eds., *Proceedings of Machine Learning Research*, Vol. 37: *International Conference on Machine Learning*, 2015, pp. 1889–1897.

Schulman, John, Filip Wolski, Prafulla Dhariwal, Alec Radford, and Oleg Klimov, "Proximal Policy Optimization Algorithms," *arXiv*: 1707.06347, last updated August 28, 2017.

Shannon, Claude E., "Programming a Computer for Playing Chess," *Philosophical Magazine*, Vol. 41, No. 314, 1950, pp. 256–275.

"'Shifter' Makes Container-Based HPC a Breeze," blog post, National Energy Research Scientific Computing Center, August 11, 2015. As of December 17, 2019: http://www.nersc.gov/news-publications/nersc-news/nersc-center-news/2015/shifter-makes-container-based-hpc-a-breeze/

Silver, David, Aja Huang, Chris J. Maddison, Arthur Guez, Laurent Sifre, George van den Driessche, Julian Schrittwieser, Ioannis Antonoglou, Veda Panneershelvam, Marc Lanctot, Sander Dieleman, Dominik Grewe, John Nham, Nal Kalchbrenner, Ilya Sutskever, Timothy Lillicrap, Madeleine Leach, Koray Kavukcuoglu, Thore Graepel, and Demis Hassabis, "Mastering the Game of Go with Deep Neural Networks and Tree Search," *Nature*, Vol. 529, January 27, 2016, pp. 484–489.

Sutton, Richard S., and Andrew G. Barto, "Chapter 6: Temporal-Difference Learning," *Reinforcement Learning: An Introduction*, 2nd ed., Cambridge, Mass.: MIT Press, 2018.

Sylabs.io, "Documentation and Examples," webpage, undated. As of December 17, 2019: https://www.sylabs.io/

Tarateta, Maja, "After Winning Jeopardy, IBM's Watson Takes on Cancer, Diabetes," *Fox Business*, October 7, 2016. As of February 17, 2020: https://www.foxbusiness.com/features/after-winning-jeopardy-ibms-watson-takes-on-cancer-diabetes

U.S. Air Force, Exercise Plan 80: RED FLAG—NELLIS (RF-N), COMACC EXPLAN 8030, ACC/A3O, July 2016a.

U.S. Air Force, F-16 Pilot Training Task List, ACC/A3TO, June 2016b, Not available to the general public.

U.S. Air Force, Office of the Chief Scientist, *Autonomous Horizons: System Autonomy in the Air Force–A Path to the Future*, Vol. 1: *Human-Autonomy Teaming*, Washington, D.C., 2015.

Vincent, James, "AI Bots Trained for 180 Years a Day to Beat Humans at Dota 2," *The Verge*, June 25, 2018. As of December 17, 2019: https://www.theverge.com/2018/6/25/17492918/openai-dota-2-bot-ai-five-5v5-matches

Weber, Bruce, "Swift and Slashing, Computer Topples Kasparov," *New York Times*, May 12, 1997.

White, Samuel G., III, *Requirements for Common Bomber Mission Planning Environment*, Wright-Patterson Air Force Base, Ohio: Department of the Air Force, Air University, Air Force Institute of Technology, AFIT/IC4/ENG/06-08, June 2006.

Wu, Yonghui, Mike Schuster, Zhifeng Chen, Quoc V. Le, Mohammad Norouzi, Wolfgang Macherey, Maxim Krikun, Yuan Cao, Qin Gao, Klaus Macherey, Jeff Klingner, Apurva Shah, Melvin Johnson, Xiaobing Liu, Łukasz Kaiser, Stephan Gouws, Yoshikiyo Kato, Taku Kudo, Hideto Kazawa, Keith Stevens, George Kurian, Nishant Patil, Wei Wang, Cliff Young, Jason Smith, Jason Riesa, Alex Rudnick, Oriol Vinyals, Greg Corrado, Macduff Hughes, and Jeffrey Dean, "Google's Neural Machine Translation System: Bridging the Gap Between Human and Machine Translation," *arXiv*: 1609.08144, last updated October 8, 2016.

About the Authors

Li Ang Zhang is an associate engineer at the RAND Corporation. His research interests include applying machine learning, optimization, and mathematical modeling toward the policy areas of defense and technology. Zhang received his Ph.D. in chemical engineering from the University of Pittsburgh.

Jia Xu, formerly a senior engineer at the RAND Corporation, is chief architect for urban air mobility systems at Airbus. His research interests include aircraft design, unmanned aerial vehicles, and autonomy/artificial intelligence. He holds a Ph.D. in aeronautics and astronautics from Stanford University.

Dara Gold is an associate mathematician at the RAND Corporation, where she is interested in mathematical modeling and simulation across a variety of applications. She holds a Ph.D. in math from Boston University, where she specialized in differential geometry.

Jeff Hagen is a senior engineer at the RAND Corporation. His research areas include technological systems analysis and strategic policy and decisionmaking. He received his M.S. in aeronautics and astronautics from the University of Washington.

Ajay K. Kochhar is a technical analyst at the RAND Corporation. He earned an M.S. degree in computer science from the University of California, Santa Barbara, and an M.S. in physics from the University of Pittsburgh.

Andrew J. Lohn is an engineer at the RAND Corporation and a professor of public policy at the Pardee RAND Graduate School. He works primarily in areas of technology policy, often within a national or global security context. He holds a Ph.D. in electrical engineering from the University of California, Santa Cruz.

Osonde A. Osoba is an information scientist at the RAND Corporation and a professor at the Pardee RAND Graduate School. He has a background in the design and optimization of machine learning algorithms. Osoba received his Ph.D. in electrical engineering from the University of Southern California.